AF593505

THE SUN

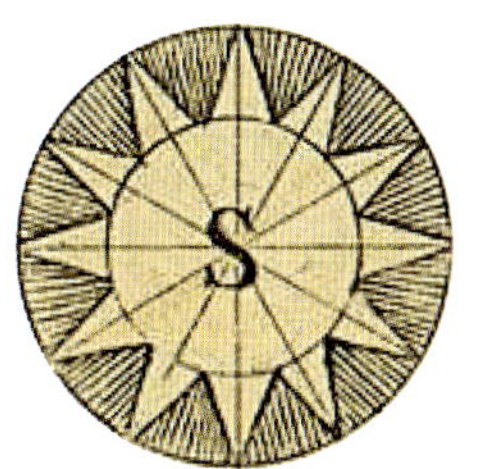

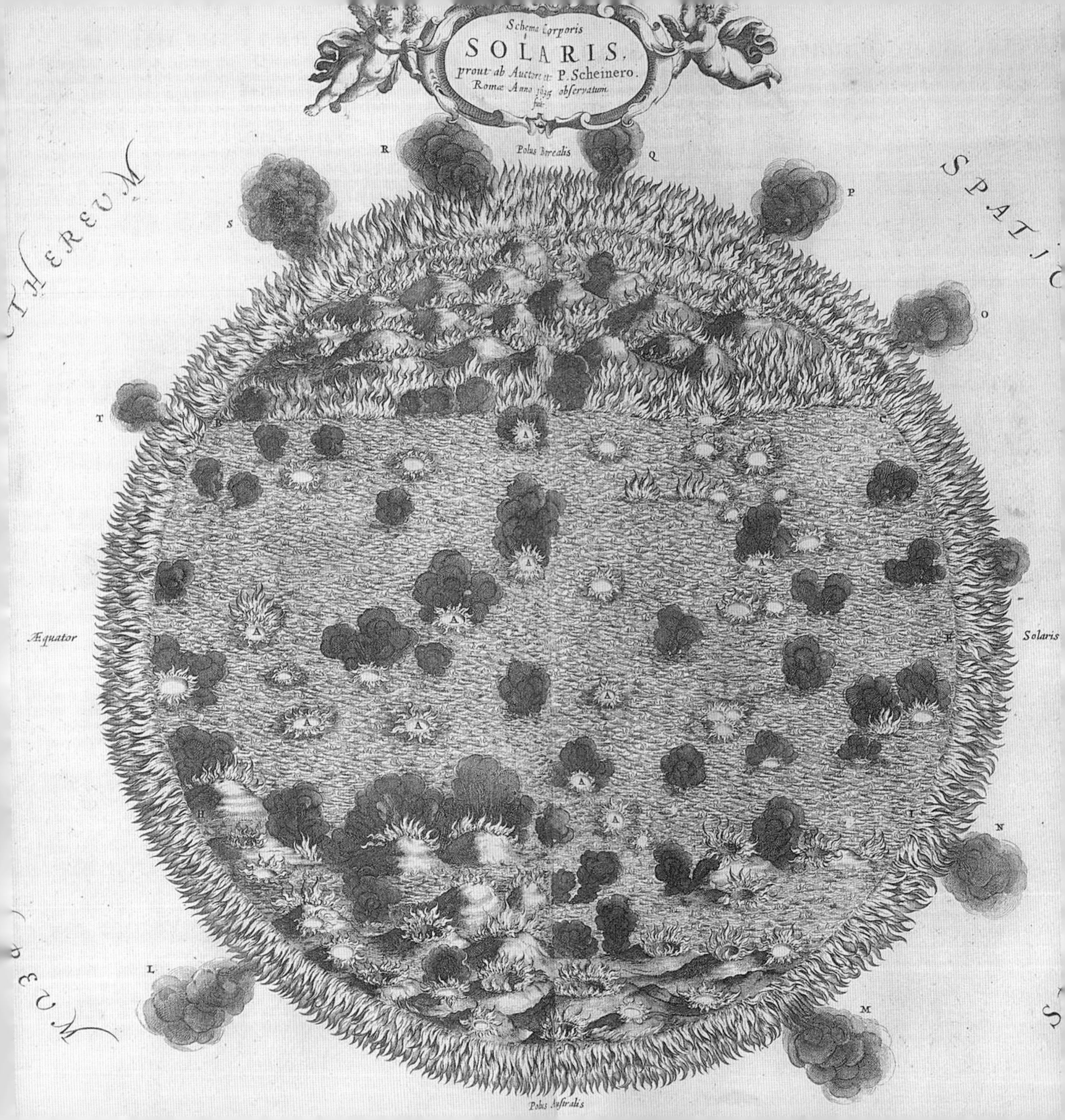
Schema Corporis
SOLARIS,
prout ab Auctore P. Scheinero.
Romæ Anno 1635 observatum
fuit.
Polus Borealis
Æquator
Solaris
Polus Australis
R
Q
S
P
O
T
B
C
D
E
H
I
N
L
M

THE SUN

ONE THOUSAND YEARS OF SCIENTIFIC IMAGERY

Katy Barrett and Harry Cliff

SCALA

SCIENCE MUSEUM

Produced exclusively for SCMG Enterprises Ltd by Scala Arts & Heritage Publishers, upon the occasion of *The Sun* exhibition at the Science Museum, London, October 2018 – May 2019.

First published in 2018 by
Scala Arts & Heritage Publishers Ltd
10 Lion Yard
Tremadoc Road
London SW4 7NQ, UK
www.scalapublishers.com

In association with
Science Museum
Exhibition Road
London SW7 2DD
www.sciencemuseum.org.uk

Every purchase supports the museum.

Project manager and copy editor: Linda Schofield
Designer: Raymonde Watkins
Printed and bound in Turkey

ISBN 978-1-78551-172-1

10 9 8 7 6 5 4 3 2 1

British Library Cataloguing in Publication Data.
A catalogue record for this book is available from the British Library.

Frontispiece: *Schema corporis solaris*, Athanasius Kircher, 1678 (see pp. 64–5).
Front and back cover: designed by Science Museum Group using elements from an engraving by J Mynde partly after a drawing of the planets orbiting the Earth by James Ferguson, 1757 (see p. 28).
Page 6: *Total Eclipse of the Sun, 22 December 1870*, Étienne Léopold Trouvelot, 1876 (see p. 107).

CONTENTS

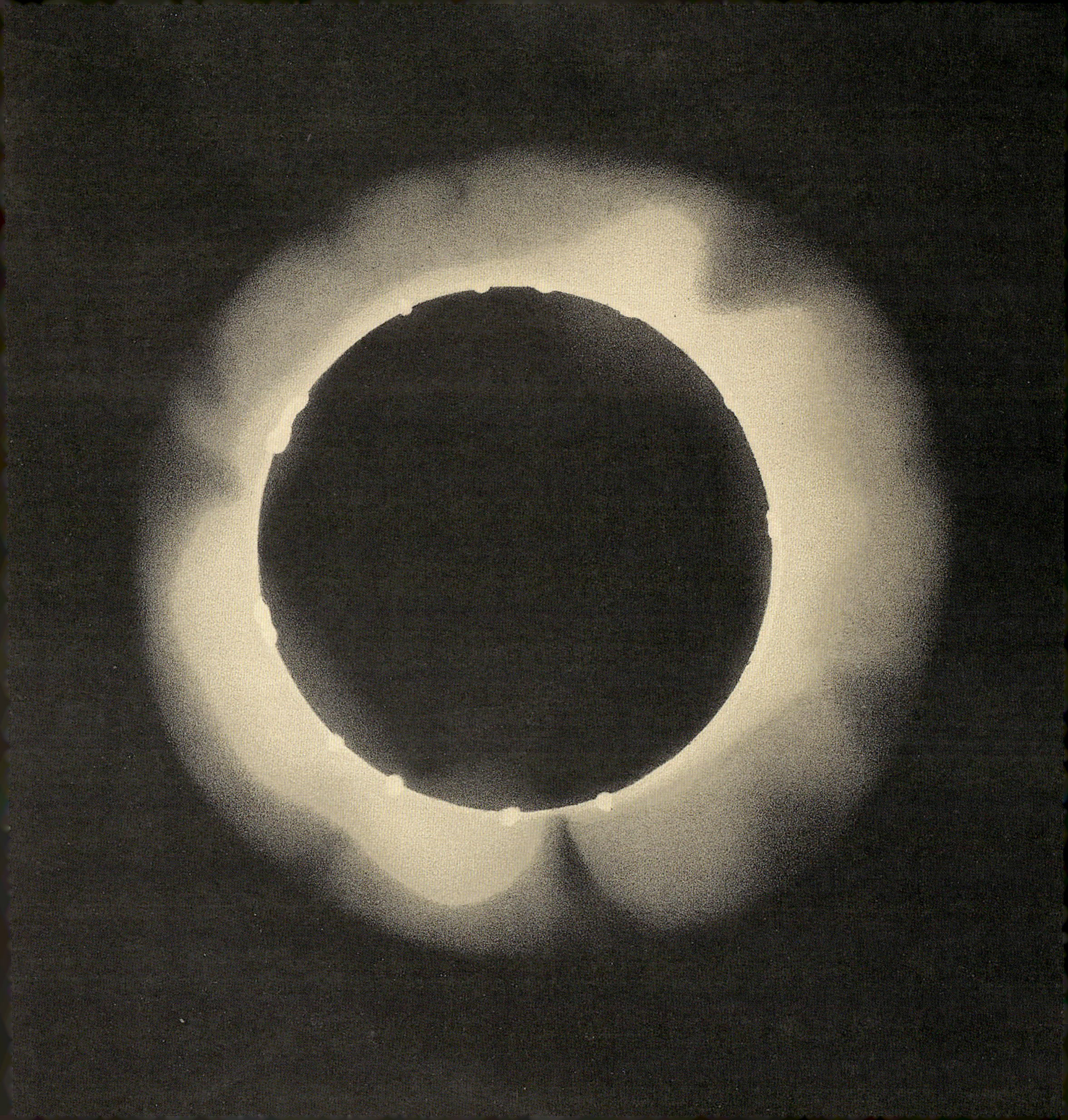

FOREWORD

The Science Museum stands on a site of extraordinary importance for Sun-watchers. In the late 19th century, it was home to the South Kensington Solar Physics Observatory, established under the leadership of Norman Lockyer to unlock the secrets of the Sun, or in his own words 'to take the very Sun to pieces'.* Famous as the codiscoverer of helium in the Sun in 1868 and as the founding editor of the English science journal *Nature*, Lockyer was also a key player in the establishment of the Science Museum, assembling a wide range of scientific instruments that became a cornerstone of our world-leading collections.

Today the Science Museum is celebrating its long-standing links with solar science with this book of beautiful and thought-provoking scientific imagery. The Museum is the custodian of one of the finest collections of solar imagery in the world, drawn particularly from Europe and the USA, from the shimmering sunspot paintings of James Nasmyth to the otherworldly photographic work of John Evershed. Through these pages you will discover a sumptuous array of sketches, paintings and photographs that reveals an ever-changing perspective on our nearest star.

Of all natural phenomena, the Sun perhaps has the greatest power to move and inspire us. It lights and warms our lives; its rising and setting shapes our days and our years. Most fundamentally of all, it gives us life. It is unsurprising that many ancient cultures worshipped the Sun, and though in our large cities it can often feel remote, the enthusiasm for eclipses shown by people worldwide is testament to its continuing hold on our imaginations.

In fact, it was a partial solar eclipse in the UK that sparked my own interest in the depiction of eclipses in Western art. Although all the images presented here were made in the course of scientific practice, be they direct observations of the Sun or attempts to communicate the latest findings, they can also be appreciated as works of art, with the power to inspire wonder, delight and awe.

The Science Museum is dedicated to engaging and inspiring the public with the science that shapes our lives and helps us to make sense of the world around us. As we celebrate the 150th anniversary of the discovery of helium, there is no better time to follow the ongoing quest to capture and understand our Sun. I would like to extend my sincere thanks to all the many partner institutions, organisations and individuals who have made this book and accompanying exhibition possible.

IAN BLATCHFORD

DIRECTOR AND CHIEF EXECUTIVE OF THE SCIENCE MUSEUM

* Iwan Rhys Morus, *When Physics Became King* (Chicago: University of Chicago Press, 2005), p. 215.

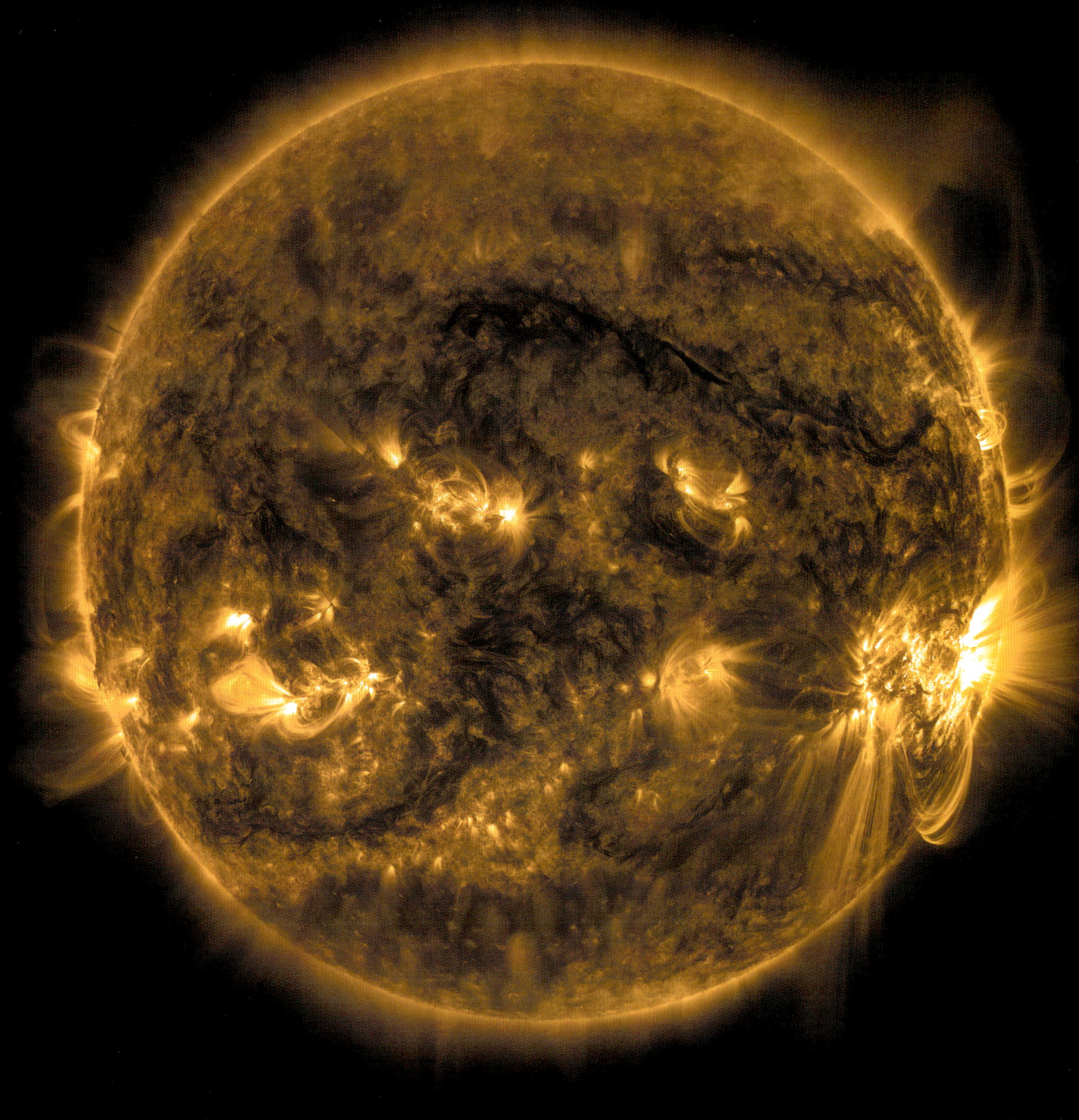

INTRODUCTION

The Sun is the star at the centre of our solar system. It is 4.5 billion years old and large enough to hold 1.3 million planet Earths within its volume. The Sun fuses hydrogen into helium at a temperature of 15 million degrees Celsius, releasing energy equivalent to 100 billion hydrogen bombs exploding every second. It will carry on doing this for another 5 billion years. It takes sunlight up to a million years to escape the core and reach the surface, but only eight minutes to travel the 150 million kilometres from the Sun to the Earth. The Sun contains 99.8% of the mass of the entire Solar System, which is held together by its gravity. On Earth, we are buffeted by the flares and eruptions that the Sun sporadically sends our way.

These are just some of the awe-inspiring facts that humans have discovered about our nearest star. But, for much of history, the Sun was a glowing enigma in our sky. The Sun is fundamental to human existence: its rising and setting shapes our daily routines, it creates the seasons, drives our weather and makes life on Earth possible. Since people first looked up at the sky the Sun has been a source of fascination, awe and inspiration. This book explores the imagery created over the last thousand years by philosophers, astronomers, artists and scientists as they strove to understand our Sun. In that time, we have moved from thinking of ourselves as living at the centre of a quiet, finite universe, where a perfect, incorruptible Sun circled the Earth, to realising we are living on the surface of an insignificant rock in orbit about a vast and violent stellar furnace, within a seemingly endless cosmos.

This transformation in our understanding of the Sun has required cultural as well as technological shifts. The images discussed here are at the centre of that story. Not simply illustrations, they are working tools that scientists have used to understand what they saw when they looked at the Sun, and to communicate their findings to each other and to an increasingly fascinated public. These pictures not only allowed them to analyse what they were seeing, but also revealed the advantages and disadvantages of different image types, from drawing to photography. Some of the scientists produced their own images; others were made for them by artists whose names we may or may not know. Some worked with assistants, scribes, instrument makers and publishers, meaning that the images and texts featured here are almost always the product of collaborative work. While the name on the image is so often that of a white male from Europe or America, we must always remember the invisible contributors who were so often female, lower class or non-Western, and hard to uncover in the histories of both science and art. The few that can be included here in their own right are therefore all the more extraordinary.

Over the past thousand years, the technologies available for both observing and imaging have changed vastly. People in the ancient and medieval worlds looked at the skies with the naked eye. The brilliance of the Sun made direct viewing impossible, of course, except when its light was filtered by passing cloud or a bank of fog. Instead, observers made inferences about the motion of the Sun and the planets through measurements of their changing positions in the heavens. They recorded their conclusions with pencil, ink and brush, on vellum, parchment or paper, and their thoughts circulated slowly through the exchange of manuscripts, both within and across the borders of Europe. The first fundamental shift came in 1440 with the invention of the printing press, which allowed multiple copies of texts and images to be disseminated more widely in ever-larger numbers.

Then, in 1608, a spectacle maker in the Netherlands submitted the first patent application for a telescope. This tube with two glass lenses, quickly adopted by astronomers, allowed users not only to discover new moons, planets and stars, but also to explore the surface of the Sun. By gathering sunlight and projecting it onto a screen, telescopes could be adapted into helioscopes, making features previously hidden by the blinding light of the Sun visible to the human eye for the first time. The solar surface was at last revealed, beginning four centuries of direct observation of the Sun.

In the 19th century, the invention of photography opened up new methods of observation and new types of imagery. Surfaces treated with photosensitive chemicals can detect more than the human eye, reaching beyond the visible spectrum and accumulating light over time, allowing people to see farther. In Britain the sensitised surface was paper, while in France images were captured on polished silver-plated copper that had been made light sensitive, and given the name daguerreotypes. The Moon was the first celestial body to feature in a successful astronomical daguerreotype taken in New York in 1840.

To photograph the Sun, the challenge was to limit rather than increase the amount of light captured, which would otherwise saturate photographic plates. In 1854, Kew Observatory near London commissioned the first instrument designed specifically to photograph the Sun: the photoheliograph. This used an ingenious spring-loaded mechanism to draw a plate with a narrow slit through the barrel of the telescope, limiting the exposure to a fraction of a second.

Previous page
Photograph showing coronal loops, taken in extreme ultraviolet by NASA's Solar Dynamics Observatory, 2012 (see pp. 80–1).

At about the same time, astronomers' attention began to turn towards information hidden in sunlight itself. In 1802, dark lines were detected in the visible spectrum of sunlight, and over the 19th century, astronomers and chemists established that these lines encoded the chemical make-up of the Sun. Its composition could now be read by comparison with spectra created in the laboratory, while print and photography offered means to capture and map these myriad dark lines in ever-greater detail. Lithography, which printed in fine detail drawings made directly on stone, grew in popularity throughout the century and brought the prospect of reliable and replicable colour into our story, making it much easier to publish spectral images.

Solar physics emerged as a discipline with the growth of spectroscopy. Invented in the 1890s, the spectroheliograph allowed the Sun to be photographed in a single wavelength of visible or invisible light, revealing a wealth of new phenomena. Solar physicists now yearned to see beyond the Earth's atmosphere, which absorbs the majority of solar radiation. The launch of Sputnik 1 by the USSR in 1957 opened up the possibility that a human observer, a telescope or a camera might be carried outside of this filtering atmosphere. Missions to record the Sun soon followed, measuring its magnetic field, outer layers, brightness and internal structure. These missions brought a new level of international collaboration to solar observation, with data and images produced by teams of men and women from different countries and backgrounds. Spacecraft built in Europe, North America, Asia and elsewhere continue to survey the Sun and its solar system, sending back digital images for scientists to analyse, manipulate and share, like never before, with the scientific community and the public.

Throughout these changes in science, technology and culture, certain aspects of the Sun have consistently fascinated observers, whether scientists or artists. This book takes a thematic approach, focusing in turn on images of the Sun's place in the universe, its complex surface, its violent and spectacular eruptions, the structure of sunlight itself and, arguably the most affecting and revealing of astronomical events, total eclipses. Each section charts imaging and observational methods over time, taking us from paper to digital, with characters, places and techniques recurring in intriguing ways.

We end with an image of a recent solar event: the total eclipse visible over the United States in the summer of 2017. Taken by a distant spacecraft looking back towards the Earth, the Sun is revealed only by its absence, as the Moon's shadow falls across the American Midwest. In that dark circle, scientists, artists and the public came together to look at our Sun for a brief moment. Some sketched and took photographs, others made scientific measurements, while thousands more just gazed upwards, joining the ranks of the solar observers who created the images in this book.

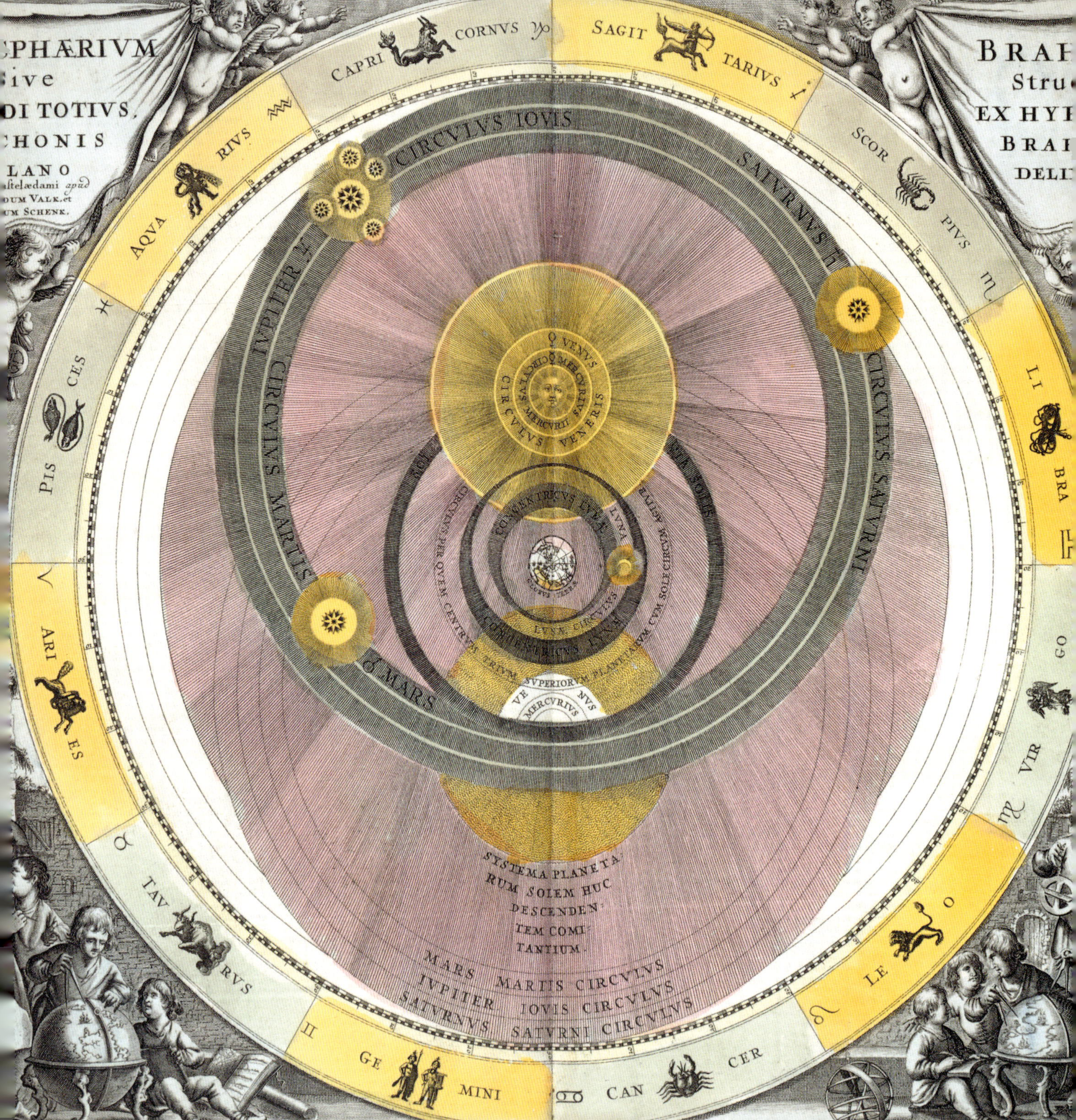
CPHÆRIVM
ive
DI TOTIVS,
CHONIS
LANO
BRA
Stru
EX HY
BRAI
DELI
CAPRI CORNVS
SAGIT TARIVS
SCOR PIVS
LI BRA
VIR GO
LE O
CAN CER
GE MINI
TAV RVS
ARI ES
PIS CES
AQVA RIVS
CIRCVLVS IOVIS
SATVRNVS
CIRCVLVS SATVRNI
IVPITER
CIRCVLVS MARTIS
MARS
VENVS
MERCVRIVS
CIRCVLVS VENERIS
CIRCVLVS MERCVRII
SOL
LVNA
LVNÆ CIRCVLVS
SVPERIORA
VE NVS
MERCVRIVS
SYSTEMA PLANETA
RVM SOLEM HVC
DESCENDEN
TEM COMI
TANTIVM.
MARS MARTIS CIRCVLVS
IVPITER IOVIS CIRCVLVS
SATVRNVS SATVRNI CIRCVLVS

1

THE SUN AND PLANETS

The Sun absolutely dominates the solar system. Almost a thousand times more massive than all the planets combined, its enormous gravitational influence holds them in orbit about the shining centre. But for half of the last millennium, the Sun was regarded as just one of the seven classical planets, albeit a particularly large and brilliant one. The Earth was the centre of God's creation, with the Sun, Moon, planets and stars carried around it once per day, embedded within invisible celestial spheres.

In the 16th century, new ideas, innovations in instrument making and an increase in the precision of astronomical measurements began to challenge this view of the heavens. The often-contentious debate over whether the Earth or the Sun sat at the centre persisted for more than a century.

Throughout this period, philosophers, theologians and astronomers produced astronomical treatises, often richly illustrated with representations of their favoured model of the universe. As astronomy grew in popularity, banners, posters and books on popular science were produced to explain the solar system to the public. With the dawn of space flight in the 20th century, cameras were launched beyond the vantage point of the Earth, capturing the revolution of the planets about their Sun in action.

EXCERPTS FROM PLINY THE ELDER'S *HISTORIA NATURALIS: DE POSITIONE ET CURSU VII PLANETARUM*

Ink on parchment by an English artist, Fleury, 900–1000
Part of Harley MS 2506
295 × 215 mm (manuscript page)
British Library

Observing the skies with the naked eye led generations of astronomers worldwide to conclude that the Earth was at the centre of the universe. The predominant early system for understanding the structure of the universe survives in a work by Claudius Ptolemy from the year 150. He presented the Sun, Moon and planets as orbiting the Earth in a series of circles, with the fixed stars in the outermost.

In this manuscript illumination from the 10th century, the artist uses the classic diagram of concentric circles that was introduced by Ptolemy's theory. The heavenly bodies are marked in red, circling on black orbits, with a crescent for the Moon and red stars for the planets and Sun, which is shown to be noticeably larger. Such manuscripts were crucial to the dissemination of astronomical knowledge in the medieval period. This text includes a discussion of Ptolemy's system by the later Roman writer Pliny the Elder and features copies of various astronomical texts. It was produced in France at the Abbey of Fleury but illustrated by an English artist.

Previous page
Detail of *Planisphaerium Braheum* by Jan van Loon, 1708 (see p. 25).

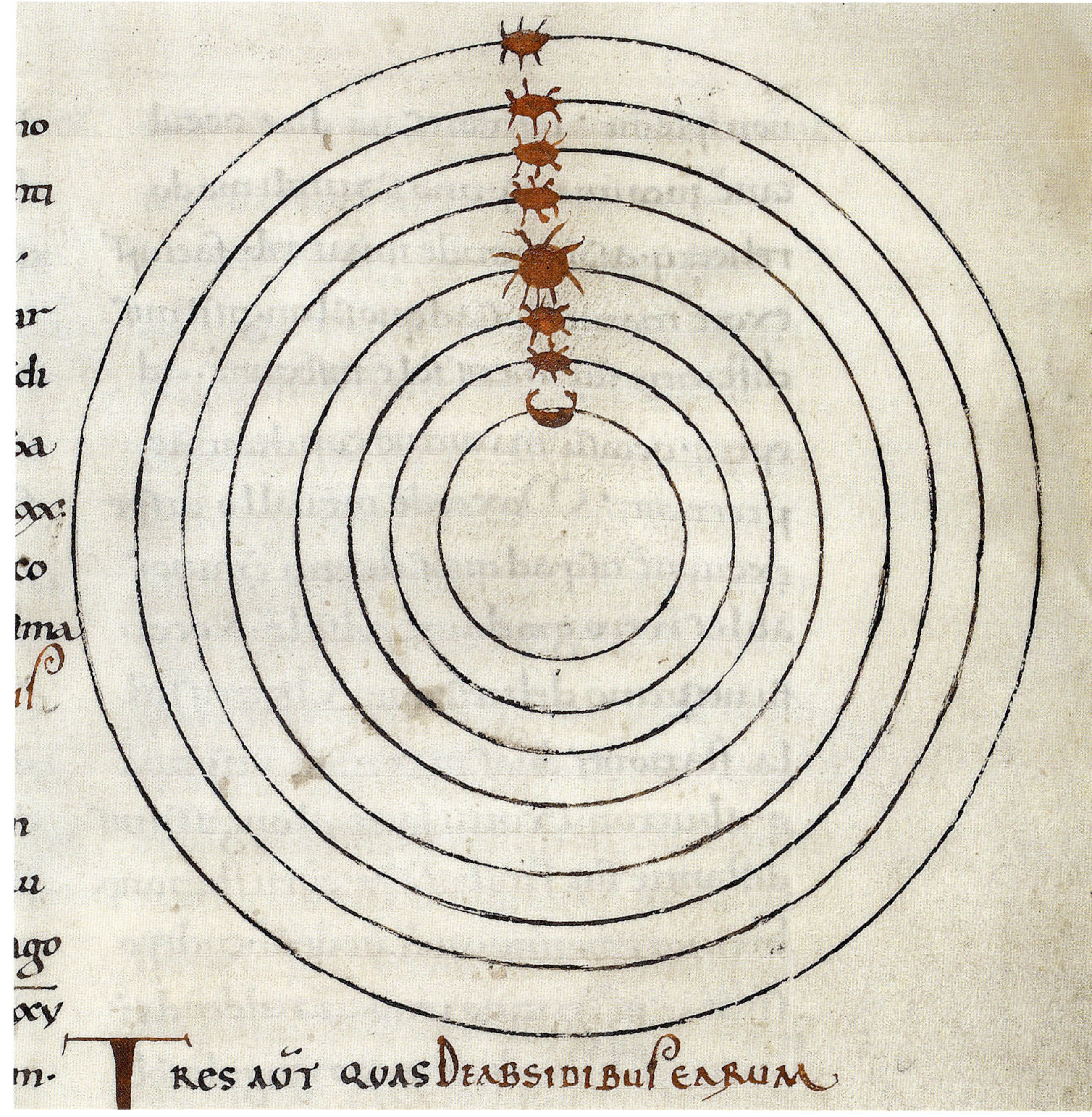
TRES AŪT QUAS DE ABSIDIBUS EARUM

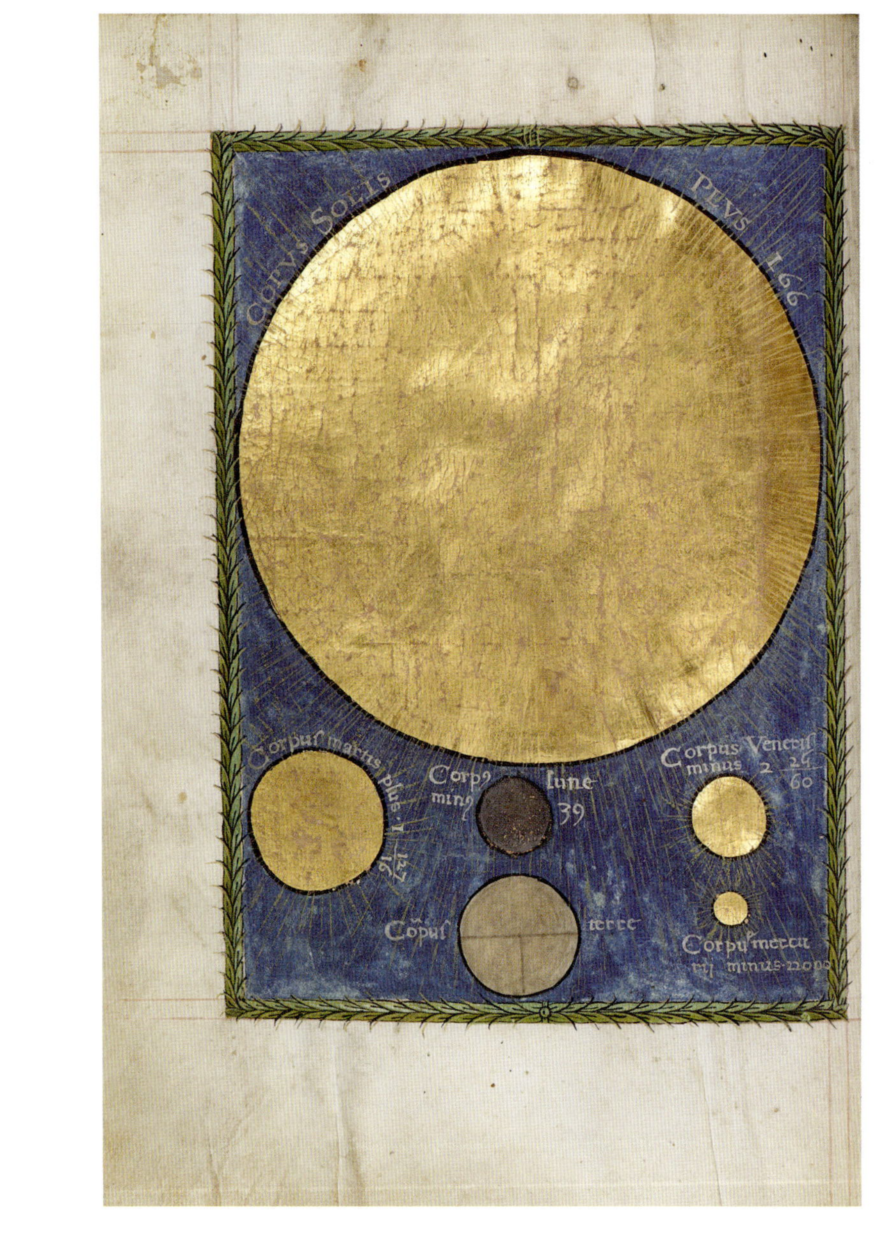
CORPVS SOLIS PLVS 166
Corpus martis plus 1 127/16
Corpus lune minus 39
Corpus Veneris minus 2 24/60
Copus terre
Corpus mercurii minus

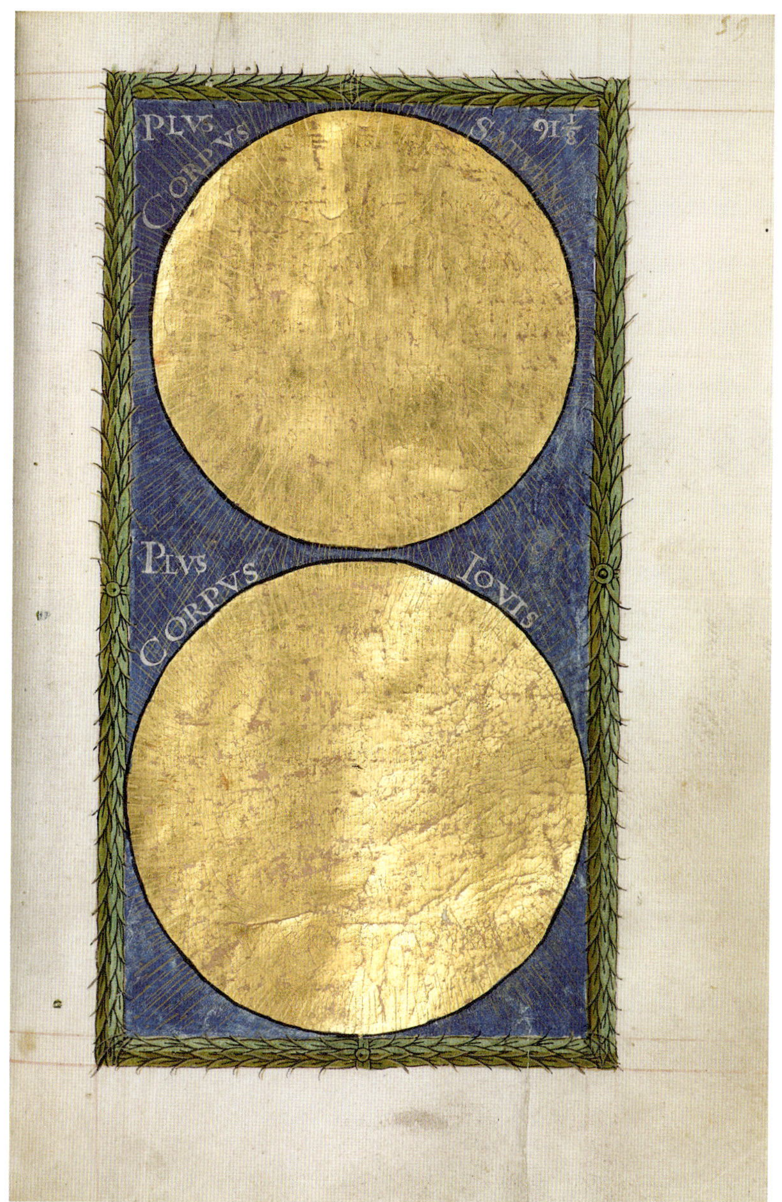

COMPARING THE SIZE OF THE SUN

Manuscript illumination on vellum by Joachinus de Gigantibus for *Astronomia* by Christianus Prolianus, Naples, 1478
212 × 140 mm each (manuscript page)
John Rylands Library

The scholar Christianus Prolianus commissioned stunningly beautiful illuminations for his book on cosmology. The work ends with this image of the sun shining in golden majesty alongside the other six known heavenly bodies and the Earth. At the time, the Sun was regarded as one of the seven planets that orbited the Earth. Accompanying text gives the volumes of the planetary bodies as multiples, or fractions, of the size of the Earth. Prolianus's universe was therefore a firmly Earth-centred system, but the Sun is clearly the largest and most important 'planet' in Joachinus de Gigantibus's illumination. In classic humanist style, Prolianus's text rehearsed the astronomical calculations of previous astronomers going back to Ptolemy and filtered through the work of Islamic philosophers like Aḥmad ibn Muḥammad ibn Kathīr al-Farghānī. Neither he nor Gigantibus had made their own observations of the sky.

For eclipse imagery from *Astronomia*, see p. 101.

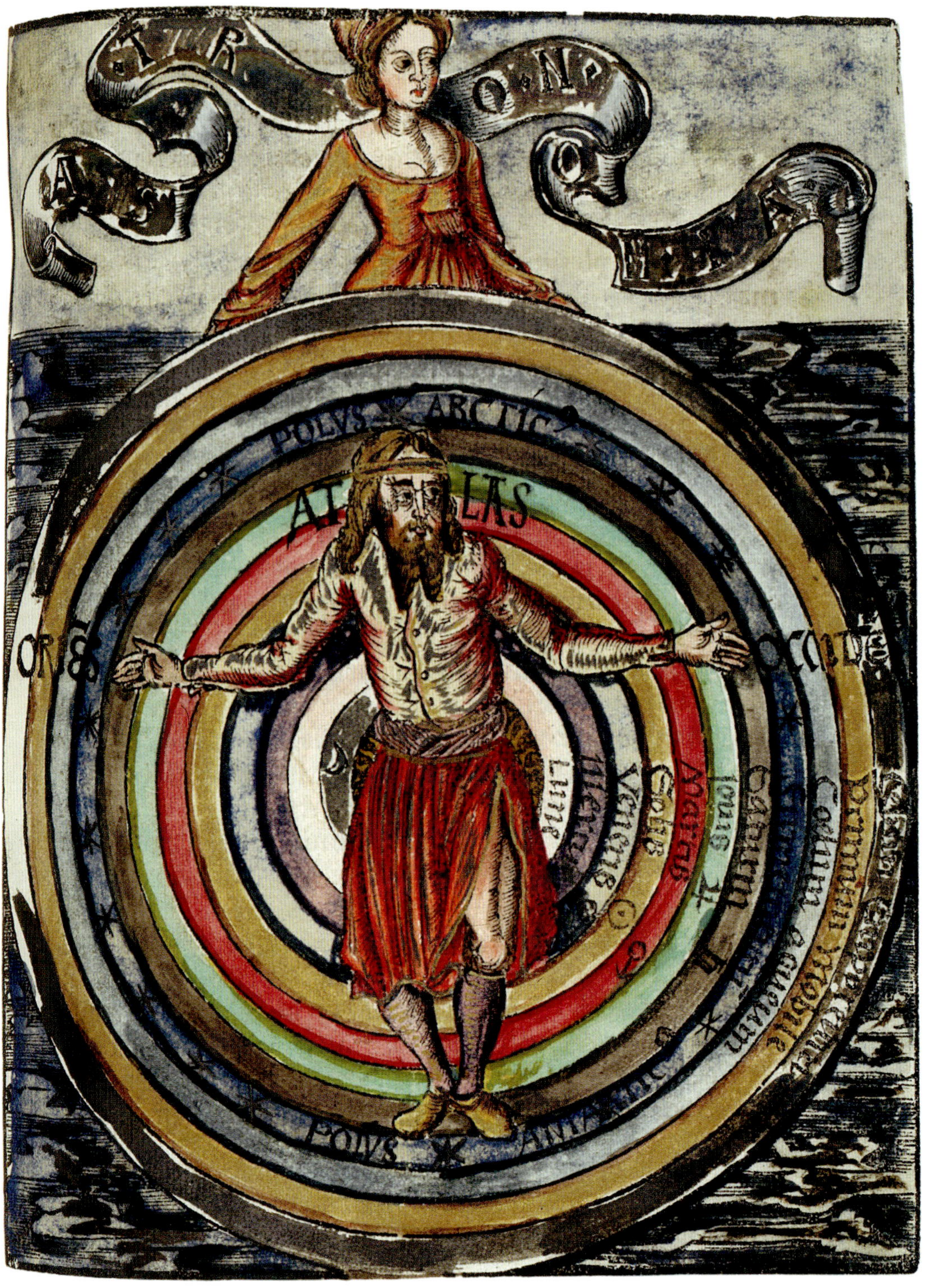
POLVS ARCTIC
ATLAS
Luna
Solis
Martis
Iouis
Saturni
Primum mobile
POLVS

ASTRONOMIA

Hand-coloured woodcut from *Margarita Philosophica* by Gregor Reisch, Basel, 1535
204 × 143 mm (manuscript page)
Science Museum Group. Object no. O.B. REI REISCH

Gregor Reisch produced one of the first encyclopedias of general knowledge, which was used widely as a university textbook in the 16th century. Astronomy featured among the 12 components of a classical education. Woodcuts accompanied each section, and here the allegorical female figure of Astronomia overlooks the Christian model of the universe. Man stands at the centre of a Ptolemaic Earth-centred system with his arms orienting the east and west, and his torso the Arctic and Antarctic poles. The outermost sphere beyond the planets is the *celum empyreum*, or God's domain outside of the physical universe. This example has been vibrantly hand-coloured, with the Sun's orbit in yellow and the divine realm in impenetrable black.

DE REVOLUTIONIBUS ORBIUM COELESTIUM

Print after the drawing by Nicolaus Copernicus,
published in Nuremberg, 1543
273 × 194 mm (manuscript page)
Science Museum Group. Object no. Q O.B. COP COPERNICUS

This small, unassuming diagram is one of the most important images in astronomical publishing. The mathematician Nicolaus Copernicus was dissatisfied with the Ptolemaic system, seeing it as both aesthetically ugly and mathematically over-complex. He simplified what had become, by the 16th century, a clunky model, and produced a system in which the planets orbited the Sun, rather than the Earth. Crucially, this explained the apparent backward motion of the planets against what was assumed were fixed stars.

Copernicus maintained Ptolemy's representation of the planets moving in concentric circles, with the fixed stars on the outermost sphere. His diagram, therefore, resembles Ptolemaic versions very closely, except for the single revolutionary word 'Sol' at the centre. It is almost identical to Copernicus's original manuscript drawing. His theory was slow to make an impact in the 16th century due to the revolutionary shift it required of humanity's understanding of its, and God's, place in the universe. It was the findings of 17th-century astronomers that started to support his theory, as seen in the kinds of detailed observations explored later in this book.

I. Stellarum Fixarum sphæra immobilis.

II. Saturnus anno. XXX. reuoluitur.

III. Iouis. XII. annorum reuolutio.

IIII. Martis bima reuolutio.

V. Telluris cum orbe lunari annua reuolutio.

terra

VI. Venus nonimestris.

VII. Mercury. LXXX. dierum.

Sol.

ORBIUM PLANETARUM DIMENSIONES, ET DISTANTIAS PER QUINQUE REGULARIA CORPORA GEOMETRICA EXHIBENS

Engraved print from *Mysterium Cosmographicum* by Johannes Kepler, Tubingen, 1597
469 × 365 mm
Science Museum Group. Object no. O.B. KEP KEPLER

Johannes Kepler produced a particularly sophisticated diagram for his model of the universe. His *Mysterium Cosmographicum* was the first published defence of the Copernican heliocentric system, but Kepler searched especially for the geometric harmony that he thought showed God's work in the universe. He constructed a new model using the five 'Platonic solids': the five regular polyhedral shapes of classical geometry in which the number of identical faces that meet at each point is the same. Kepler argued that these explained the distances between the known planets.

His diagram is therefore three-dimensional to show the sphere of Saturn around the cube of Jupiter, and so on down to Mercury's octahedron. He placed the whole model on an elaborate decorative foot, as if the universe itself were an expensive, bronze, scientific instrument.

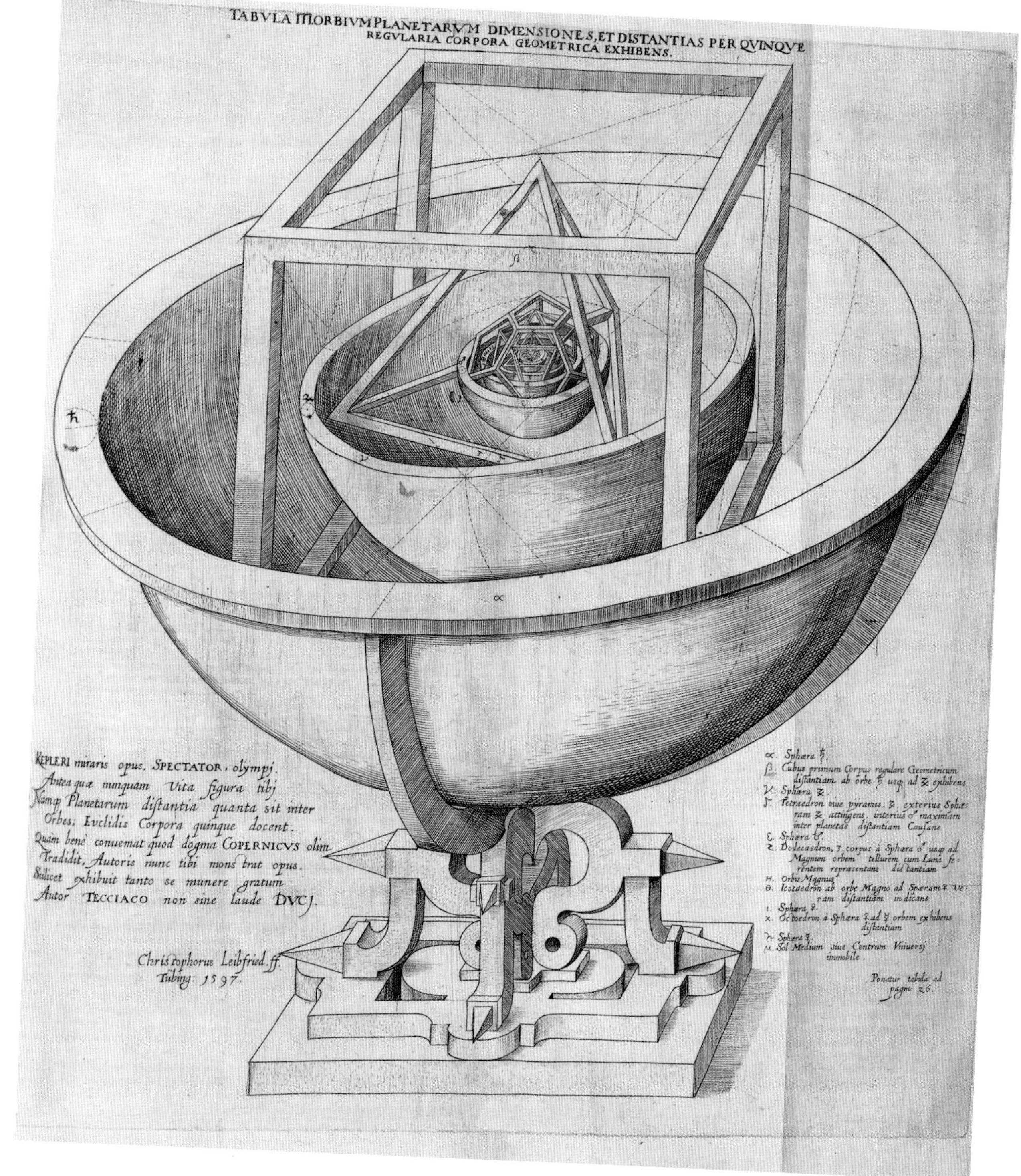
TABVLA III ORBIVM PLANETARVM DIMENSIONES, ET DISTANTIAS PER QVINQVE REGVLARIA CORPORA GEOMETRICA EXHIBENS.
KEPLERI miraris opus, SPECTATOR, olympj.
Antea quæ nunquam Vita figura tibj
Namq; Planetarum distantia quanta sit inter
Orbes; Euclidis Corpora quinque docent.
Quam benè conuemat quod dogma COPERNICUS olim
Tradidit, Autoris nunc tibi monstrat opus.
Silicet exhibuit tanto se munere gratum
Autor TECCIACO non sine laude DVCJ.
Christophorus Leibfried. ff.
Tübing: 1597.
α. Sphæra ♄.
β. Cubus primum Corpus regulare Geometricum distantiam ab orbe ♄ usq; ad ♃ exhibens
γ. Sphæra ♃.
δ. Tetraedron siue pyramis. 3. exterius Sphæram ♃ attingens, interius ♂ maximam inter planetas distantiam Causans
ε. Sphæra ♂.
ζ. Dodecaedron, 3 corpus à Sphæra ♂ usq; ad Magnum orbem tellurem cum Luna ferentem repræsentans distantiam
η. Orbis Magnus
θ. Icosaedron ab orbe Magno ad Spæram ♀ Veram distantiam indicans
ι. Sphæra ♀.
κ. Octoedron à Sphæra ♀ ad ☿ orbem exhibens distantiam
λ. Sphæra ☿.
μ. Sol Medium siue Centrum Vniuersj immobile.
Ponatur tabula ad pagin. 26.

PLANISPHAERIUM BRAHEUM

Hand-coloured engraving by Jan van Loon from Andreas Cellarius's *Harmonia Macrocosmica*, Amsterdam, 1708
310 × 553 mm (manuscript double page)
Science Photo Library

Andreas Cellarius was both author and chief designer of the *Harmonia Macrocosmica*, one of the most beautiful celestial atlases ever produced. Of 29 sumptuously engraved plates, 21 provided a summary of different models of the universe. Each model is shown within a circle at the centre, with cherubs unfurling banners above, and astronomers with instruments below. Originally published in 1660, this edition with more subdued hand-colouring was produced by the Amsterdam publishers Gerard Valk and Petrus Schenk in 1708. They have added their names to the title cartouche.

This plate shows the planetary model of the Danish astronomer Tycho Brahe published in 1588. Brahe had proved that the planets could not move on 'crystalline spheres' made of a solid, clear, crystal-like substance, as the paths of comets would need to break through these, but he maintained an Earth-centred universe. His model therefore shows all of the planets orbiting the Sun, while the Sun and Moon orbit the Earth.

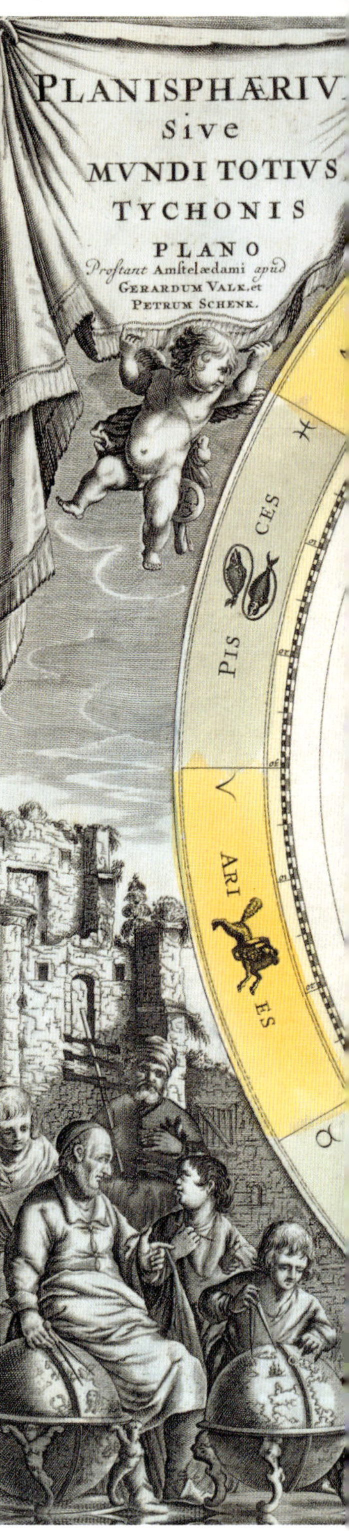

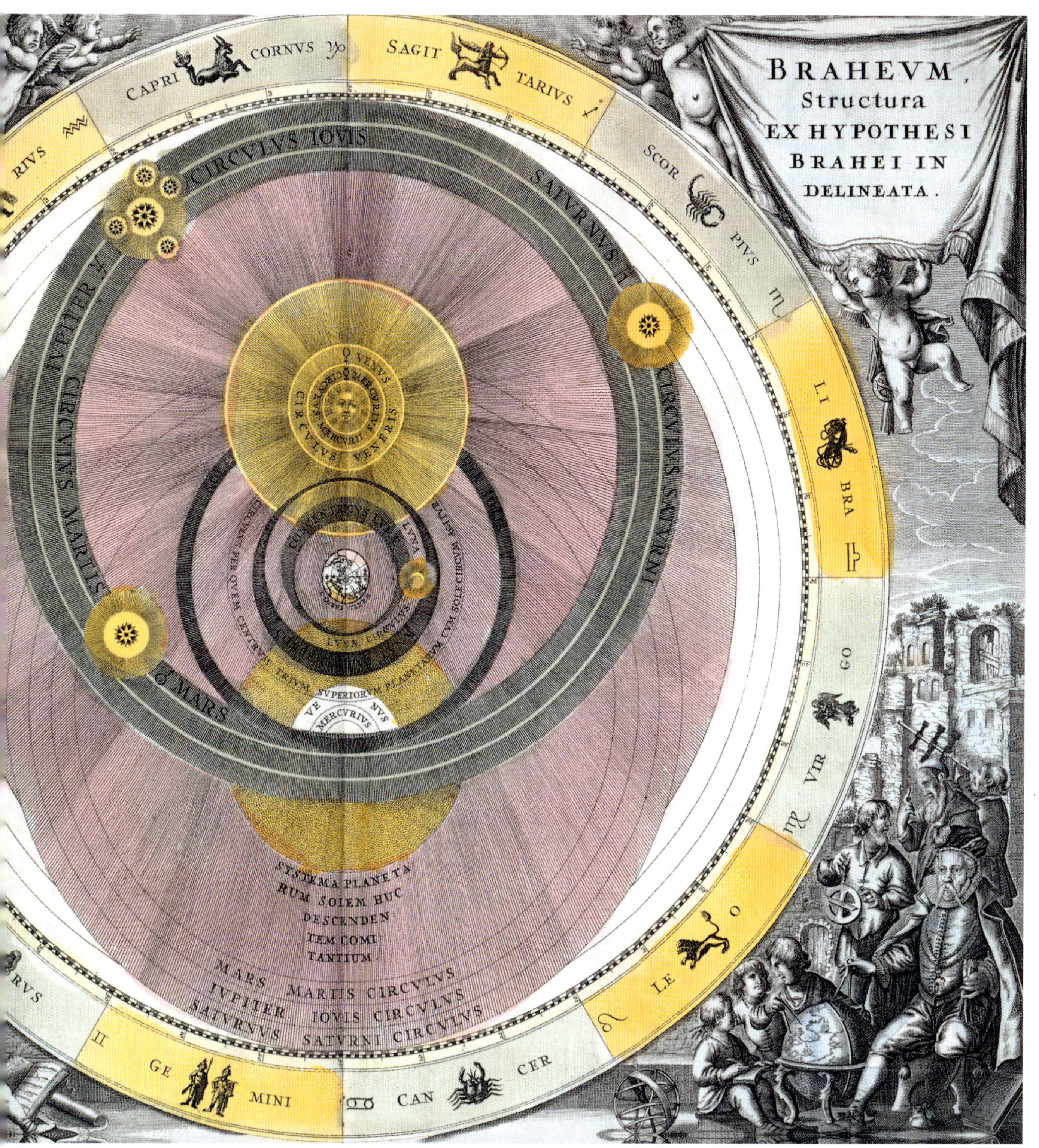
BRAHEVM, Structura EX HYPOTHESI BRAHEI IN DELINEATA.
SYSTEMA PLANETARUM SOLEM HUC DESCENDENTEM COMITANTIUM.
MARS MARTIS CIRCVLVS
IVPITER IOVIS CIRCVLVS
SATVRNVS SATVRNI CIRCVLVS

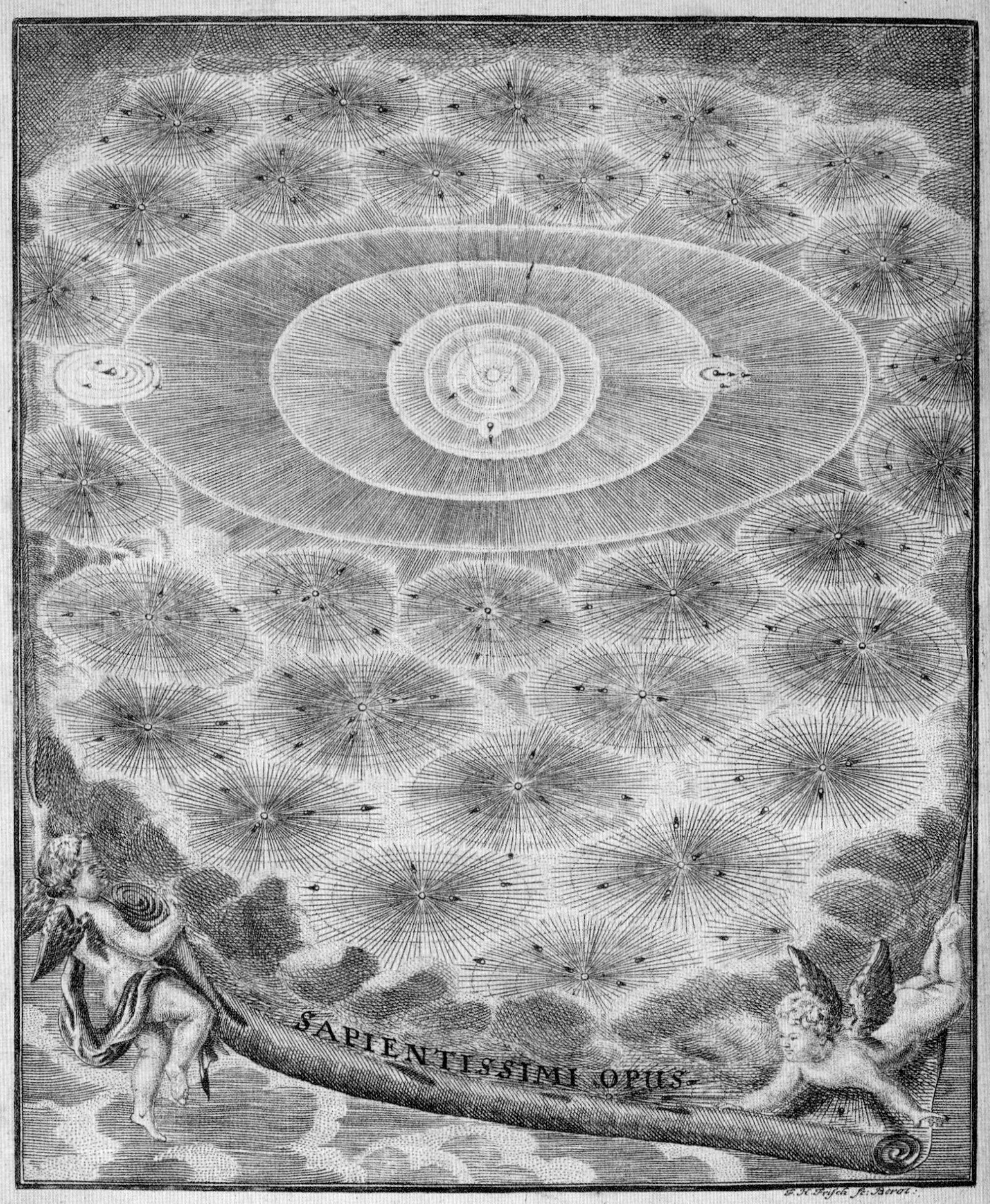

G. H. Frisch sc: Berol:

A PLURALITY OF WORLDS

Engraved frontispiece by Berol after F H Fritsch from Leonhard Euler's *Theoria Motuum Planetarum et Cometarum*, Berlin, 1744
180 × 155 mm (manuscript page)
British Library

The breaking of the Ptolemaic world system meant that astronomers could start to think about the universe differently. If the Earth was not the centre, maybe there was no centre: stars could become suns, perhaps with their own orbiting planets. Likewise, if the planets' motions were not due to crystalline spheres, then what was the motive force in the universe? These considerations led, inevitably, to the potential of multiple solar systems, multiple worlds and even other life forms, ideas that have inspired science and science fiction ever since. Leonhard Euler sought to explain the planets' orbits as caused by an all-pervasive fluid with possible magnetic properties. His charming illustration shows the Copernican system surrounded by many other solar systems in a universe that continues out of the frame of the image. At the bottom corners two angels unfurl the fabric of the universe.

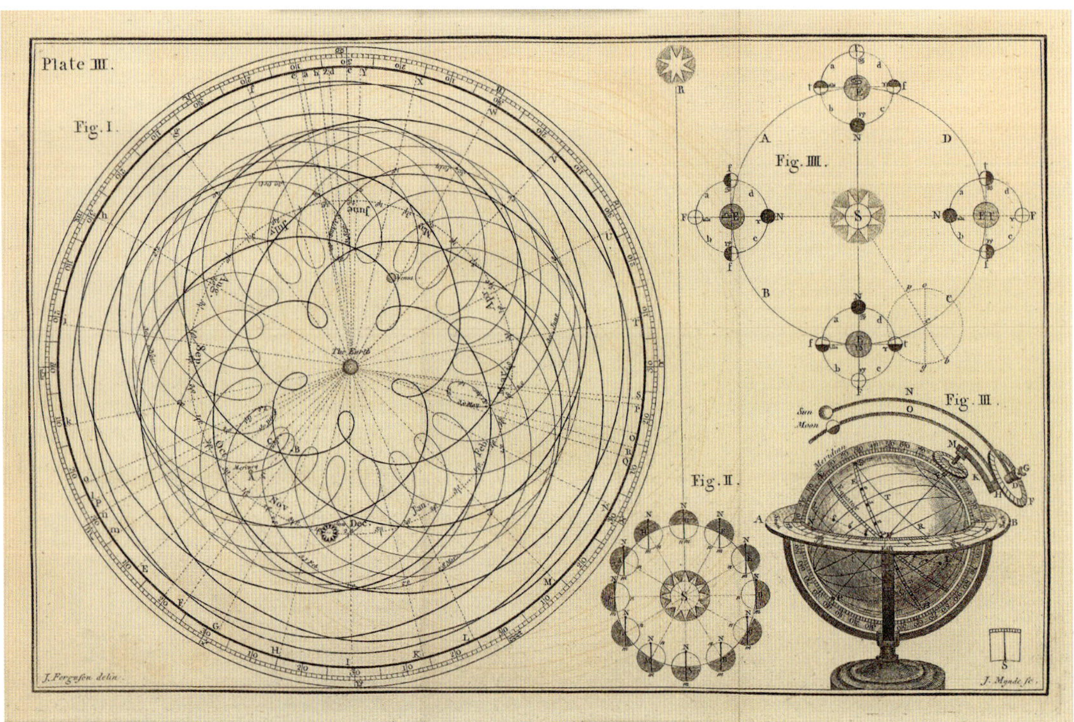

A DRAWING OF THE PLANETS ORBITING THE EARTH

Engraved by J Mynde partly after a drawing by James Ferguson published in his *Astronomy Explained upon Sir Isaac Newton's Principles*, London, 1757
258 × 400 mm
Science Museum Group. Object no. Q O.B. FER FERGUSON

James Ferguson became famous in London for his compelling lectures, illustrated by ingenious diagrams and instruments. This elegant, engraved plate would have accompanied his lectures, and was produced as a print for his first publication, which described astronomical phenomena in language familiar to ordinary readers. It was an instant and enduring success.

In this diagram, Ferguson shows the complex orbits that the planets are required to follow in a universe with the Earth at its centre. The accompanying text describes how he used an orrery – a mechanical model of the solar system – to make the original drawing. He replaced the Sun, Mercury and Venus with lead pencils, held a sheet of pasteboard over the Earth and then wound the orrery to move the pencils in the Sun and planets' places. This produced the dotted circle of months for the Sun, while Mercury and Venus drew a series of loops.

The rest of the plate uses diagrams and an illustration of an instrument to explain the solar year and harvest moons.

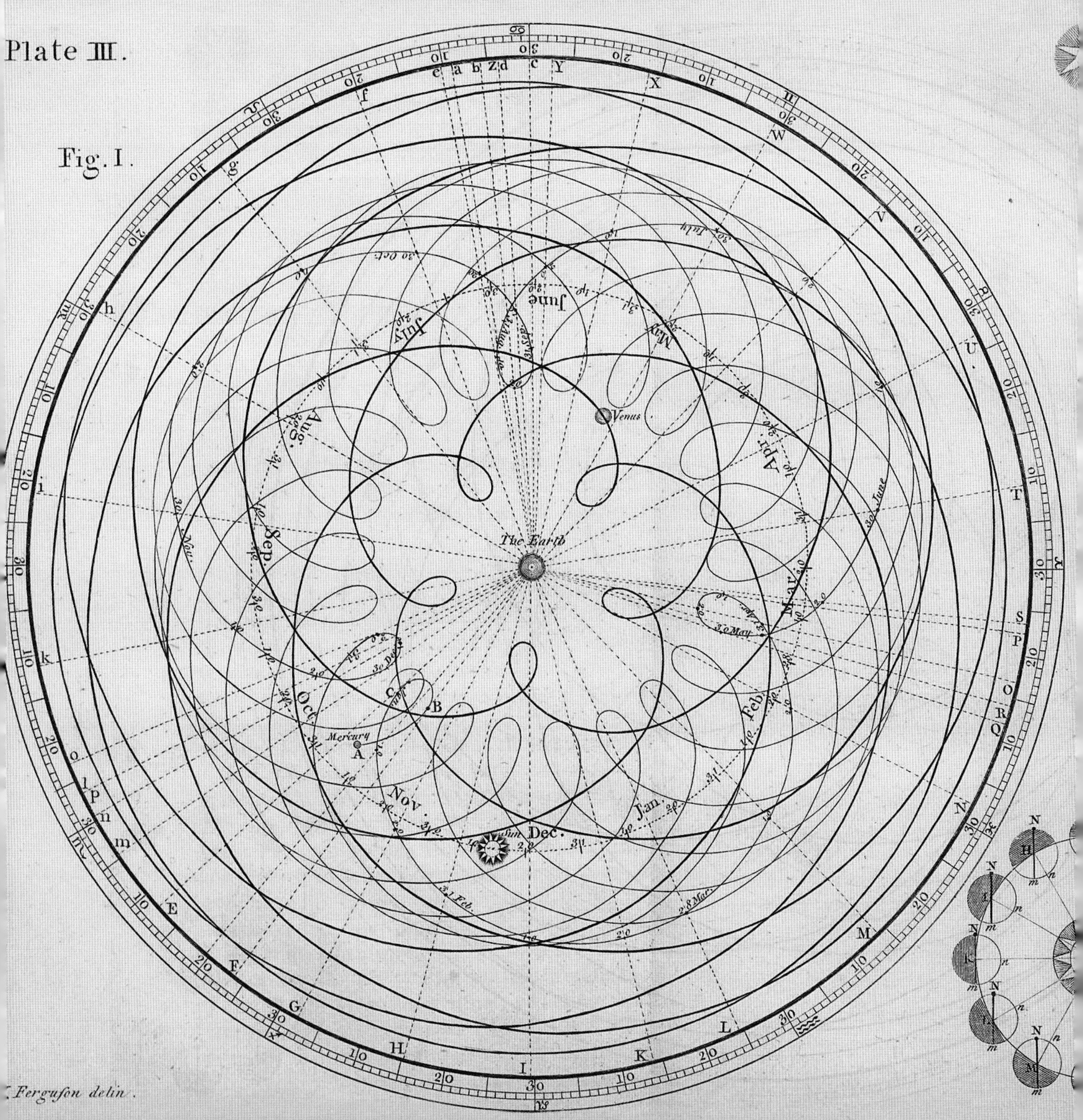
Plate III.
Fig. I.
The Earth
Venus
Mercury
Sun
Ferguson delin.

TRANSPARENT SOLAR SYSTEM,

DISPLAYING THE PLANETS WITH THEIR ORBITS, AS KNOWN AT THE PRESENT DAY.

DERIVED FROM THE LATEST AND BEST AUTHORITIES.

THE DIAMETER OF THE SUN IS 883,000 MILES; AND ITS BULK FIVE HUNDRED TIMES GREATER THAN THE WHOLE OF THE PLANETS UNITED.

THE PLANETS ARE RETAINED IN THEIR ORBITS BY THE OPPOSING ACTIONS OF THE CENTRIFUGAL AND CENTRIPETAL FORCES.

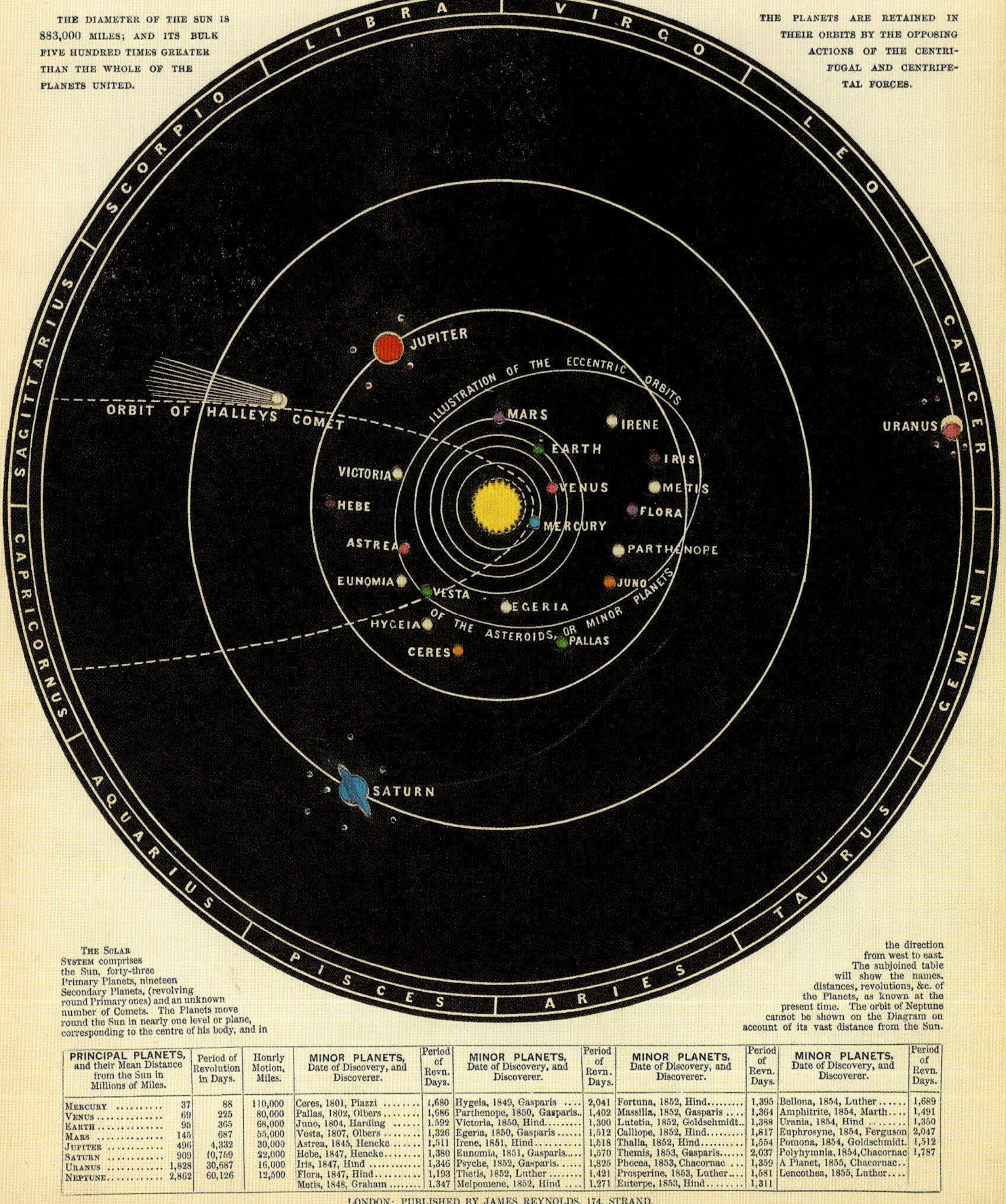

THE SOLAR SYSTEM comprises the Sun, forty-three Primary Planets, nineteen Secondary Planets, (revolving round Primary ones) and an unknown number of Comets. The Planets move round the Sun in nearly one level or plane, corresponding to the centre of his body, and in the direction from west to east. The subjoined table will show the names, distances, revolutions, &c. of the Planets, as known at the present time. The orbit of Neptune cannot be shown on the Diagram on account of its vast distance from the Sun.

PRINCIPAL PLANETS, and their Mean Distance from the Sun in Millions of Miles.		Period of Revolution in Days.	Hourly Motion, Miles.	MINOR PLANETS, Date of Discovery, and Discoverer.	Period of Revn. Days.	MINOR PLANETS, Date of Discovery, and Discoverer.	Period of Revn. Days.	MINOR PLANETS, Date of Discovery, and Discoverer.	Period of Revn. Days.	MINOR PLANETS, Date of Discovery, and Discoverer.	Period of Revn. Days.
MERCURY	37	88	110,000	Ceres, 1801, Piazzi	1,680	Hygeia, 1849, Gasparis	2,041	Fortuna, 1852, Hind........	1,395	Bellona, 1854, Luther	1,689
VENUS	69	225	80,000	Pallas, 1802, Olbers	1,686	Parthenope, 1850, Gasparis..	1,402	Massilia, 1852, Gasparis	1,364	Amphitrite, 1854, Marth....	1,491
EARTH	95	365	68,000	Juno, 1804, Harding	1,592	Victoria, 1850, Hind.........	1,300	Lutetia, 1852, Goldschmidt..	1,388	Urania, 1854, Hind	1,350
MARS	145	687	55,000	Vesta, 1807, Olbers	1,326	Egeria, 1850, Gasparis	1,512	Calliope, 1852, Hind.........	1,817	Euphrosyne, 1854, Ferguson	2,047
JUPITER	496	4,332	30,000	Astrea, 1845, Hencke	1,511	Irene, 1851, Hind..........	1,518	Thalia, 1852, Hind..........	1,554	Pomona, 1854, Goldschmidt.	1,512
SATURN	909	10,759	22,000	Hebe, 1847, Hencke.........	1,380	Eunomia, 1851, Gasparis.....	1,570	Themis, 1853, Gasparis......	2,037	Polyhymnia, 1854, Chacornac	1,787
URANUS	1,828	30,687	16,000	Iris, 1847, Hind	1,346	Psyche, 1852, Gasparis.	1,825	Phocea, 1853, Chacornac ...	1,359	A Planet, 1855, Chacornac..	
NEPTUNE...........	2,862	60,126	12,500	Flora, 1847, Hind..........	1,193	Thetis, 1852, Luther	1,421	Prosperine, 1853, Luther....	1,581	Lencothea, 1855, Luther....	
				Metis, 1848, Graham	1,347	Melpomene, 1852, Hind	1,271	Euterpe, 1853, Hind........	1,311		

LONDON: PUBLISHED BY JAMES REYNOLDS, 174, STRAND.

TRANSPARENT SOLAR SYSTEM

Drawn and engraved by John Emslie, published by James Reynolds, London, 1851
285 × 228 mm
Science Museum Group. Object no. 1987-889/3

In the 19th century, the publisher James Reynolds and the illustrator John Emslie produced a series of sets of instructional prints: forerunners of modern infographics. Their set of 12 *Astronomical Diagrams* included depictions of the solar system alongside demonstrations of the various sizes of the planets, a chart of the heavens, and illustrations of eclipses, the Earth's seasons and comets. This wonderful transparent system is backed with coloured tissue so that it can be held up to the light to make the stars and planets shine. It was inscribed as a gift 'with Uncle and Aunt Speck's best love' in April 1857.

For another print from the *Astronomical Diagrams*, see p. 51.

SOLAR SYSTEM QUILT

Cotton and wool quilt with wool-fabric appliqué, wool braid, and wool and silk embroidery by Ellen Harding Baker, Iowa, USA, 1876
2250 × 2690 mm
National Museum of American History, Kenneth E Behring Center

This quilt is a rare surviving example of female engagement with astronomy in the 19th century. Ellen Harding Baker combined a popular domestic activity – quilting – with contemporary fascination for astronomy. Study of the skies was deemed appropriate for 19th-century women and was sometimes included in their education.

Harding Baker's striking quilt features the imagery seen in astronomy textbooks of the time. The central flaming Sun has a comet looping around it and then the four closest planets in tight orbits; a tiny Moon accompanies the Earth. The four largest planets sit further out, beyond the bright colours of the asteroid belts. Harding Baker used this quilt in astronomy lectures that she gave to her neighbours in the Iowa towns of West Branch, Moscow and Lone Tree. Her name, the date and the quilt's subject are added to the corners.

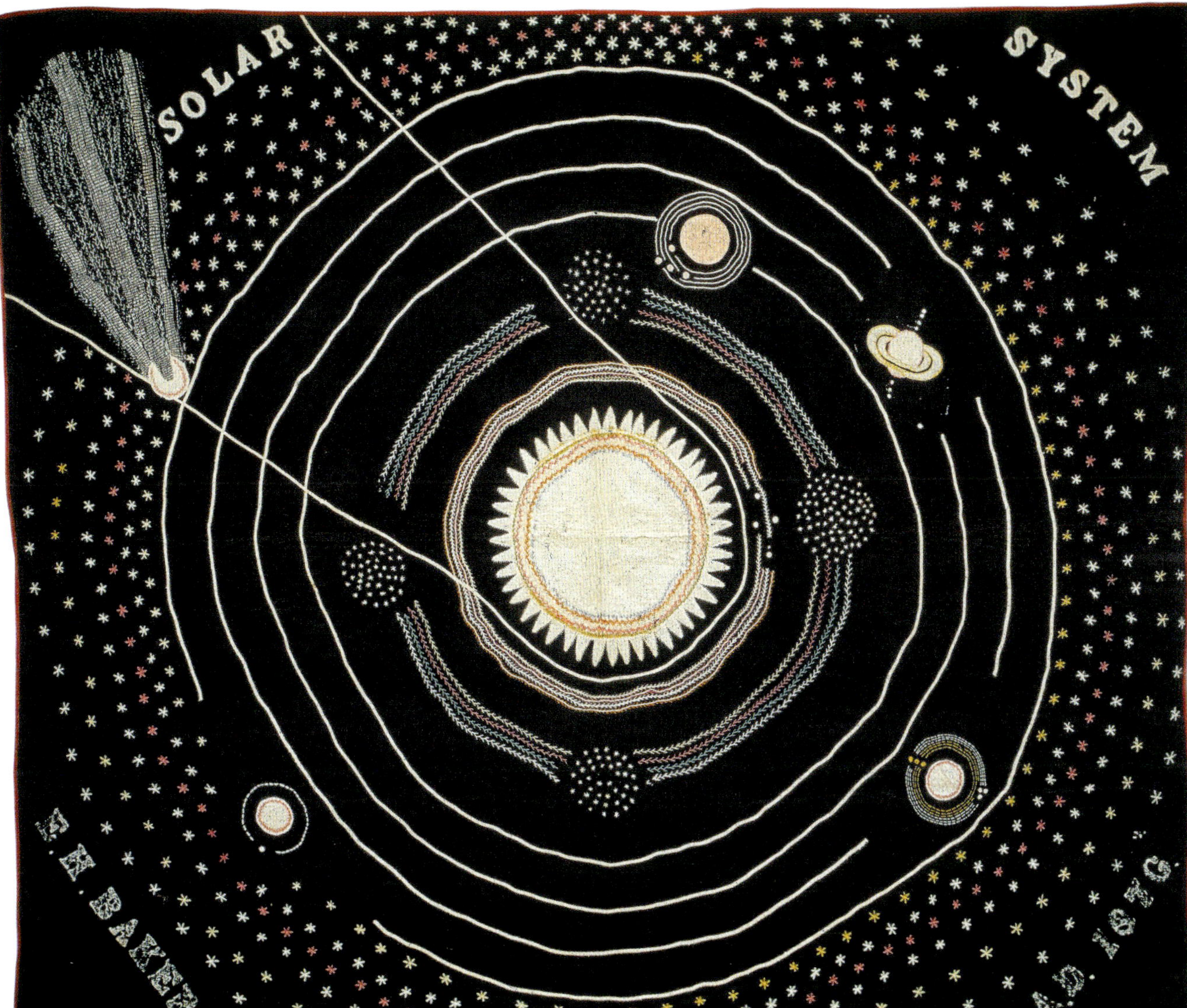
SOLAR SYSTEM
E. H. BAKER
A.D. 1876.

THE SUN AND THE PALE BLUE DOT

Composite photograph by the Voyager 1 space probe,
4 billion kilometres from Earth, 14 February 1990
Digital image
NASA

This is the most distant photograph of our Sun ever taken, captured by NASA's (National Aeronautics and Space Administration) Voyager 1 space probe as it hurtled out of the solar system. From a vantage point far beyond the orbit of Pluto, the Sun appears 26 times smaller than it does from Earth. To the left are two superimposed images containing Venus and Earth, the latter a barely visible pale blue dot hanging in the vastness of space. This image inspired the astronomer Carl Sagan to reflect that 'every saint and sinner in the history of our species, lived there – on a mote of dust, suspended in a sunbeam'.*

* From a public lecture given by Carl Sagan at Cornell University, 13 October 1994.

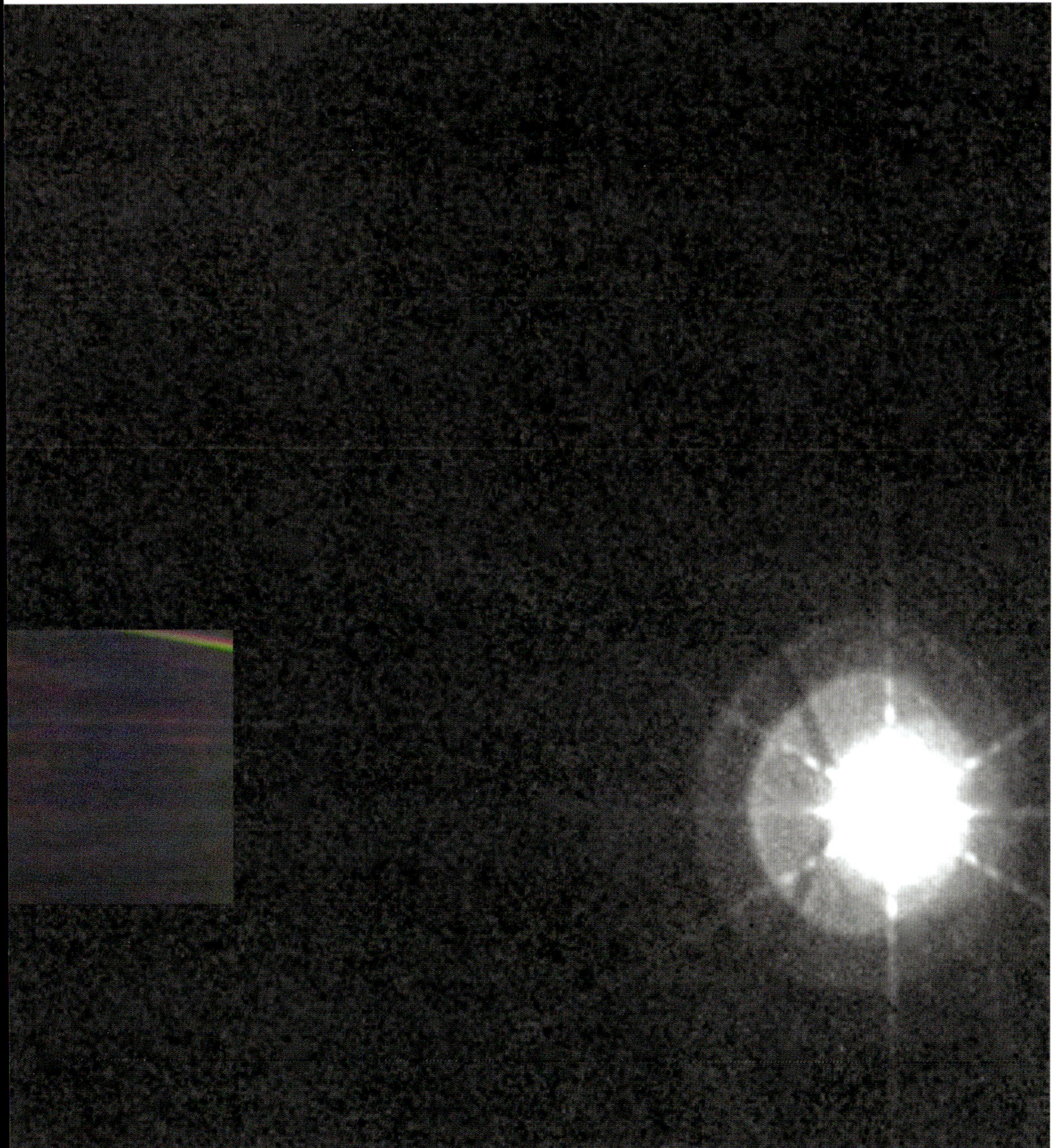

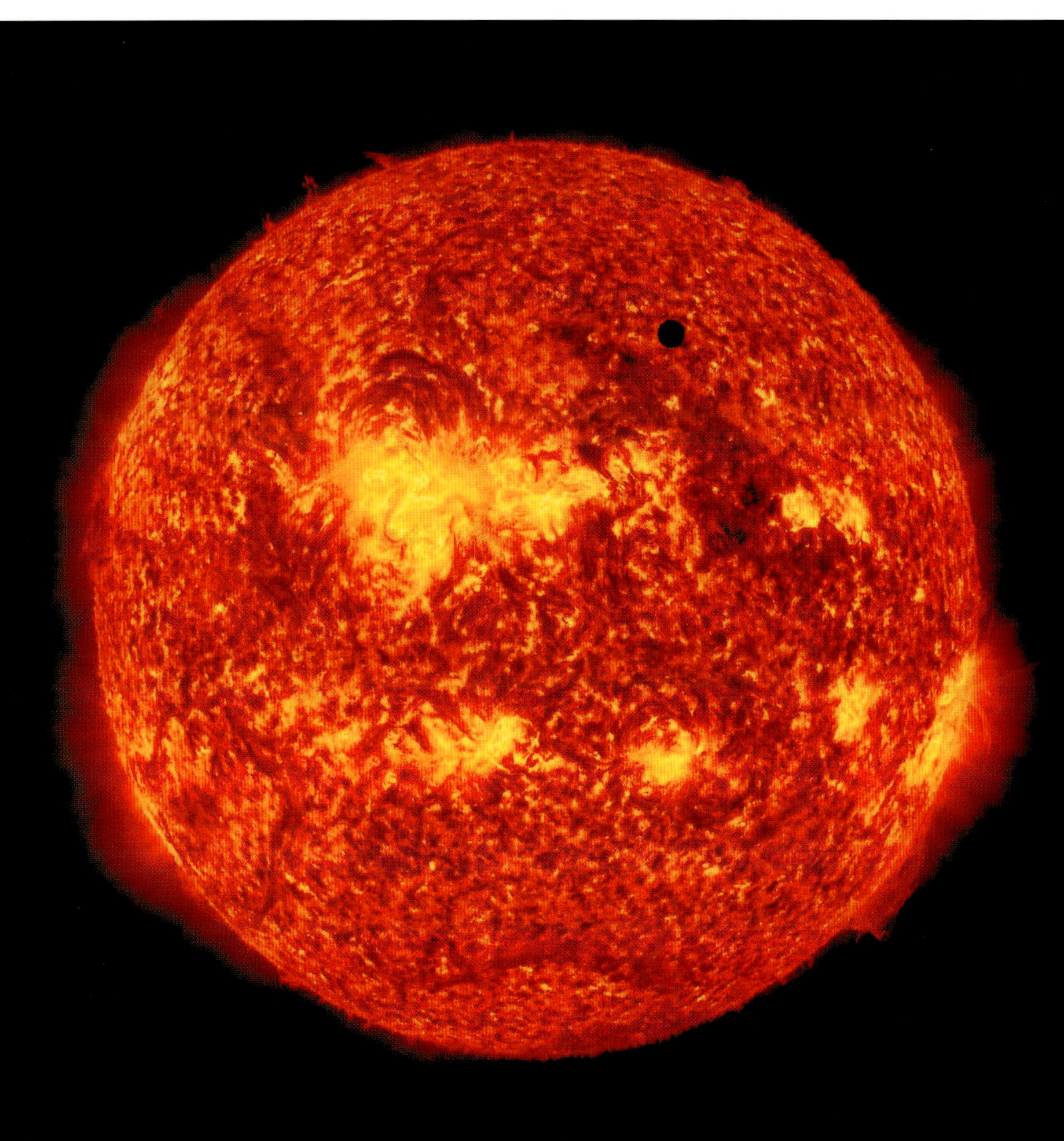

TRANSIT OF VENUS

Photograph in extreme ultraviolet light by NASA's Solar Dynamics Observatory (SDO), 5 June 2012
Digital image
NASA

In June 2012, NASA's Solar Dynamics Observatory recorded one of the rarest of all astronomical events: the passage of Venus across the face of the Sun. Imaged here in extreme ultraviolet light and artificially recoloured in flaming red, the Sun looms monstrously behind the tiny planet. It is a striking illustration of the vast scale of our Sun compared to its orbiting planets.

The chance to view a transit of Venus comes twice in a lifetime. They occur in pairs eight years apart, separated by long gaps of 105.5 or 125.5 years. The transit prior to 2012 took place in 2004, but it will be a long wait until the next chance to see one comes around again, in 2117.

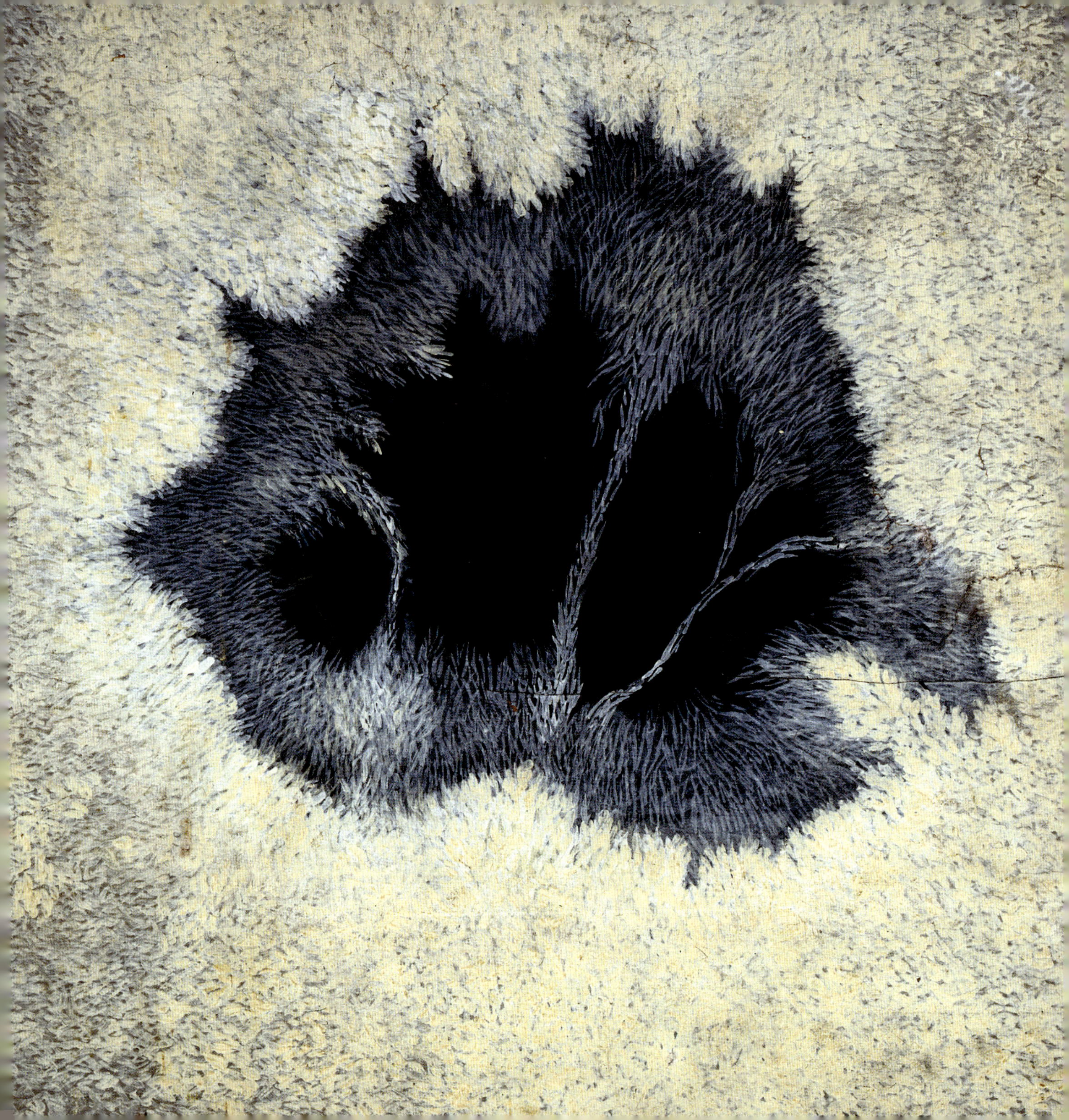

2

THE SOLAR SURFACE

For much of history, the Sun's secrets were hidden behind its dazzling glare. When people looked to the sky they saw a brilliant light, reduced at dawn and dusk to a glowing orange disc, beautiful but featureless.

The invention of the telescope changed everything. At the start of the 17th century, astronomers turned this new instrument towards the Sun, discovering it to be far from the perfect aetherial sphere they had imagined. Instead, they uncovered a dynamic body, pockmarked with mysterious dark spots, triggering debate over their nature.

As telescopes improved, the Sun's surface was revealed in ever-greater and more glorious detail. Astronomers found that they needed to become artists to capture its otherworldly features, and a whole generation of Sun-spotters began to make daily records of the number and appearance of sunspots. Later, photography allowed scientists to produce exposures of the solar surface, and yet the hyper-realistic photographs made by modern solar observatories often echo the diligent and tenacious work of the 19th-century Sun-painters.

CHRONICLE OF ENGLAND

Page from an illuminated manuscript compiled by John of Worcester with unknown scribes and artists, Worcester, by 1140
Approx. 330 × 250 mm (manuscript page)
Corpus Christi College, Oxford

This striking image, part illuminated manuscript and part diagram, is the earliest-known visual record of a sunspot. John was a monk at the Benedictine Priory of Worcester Cathedral, and the manuscript he produced there is one of the most important histories written in England in the 12th century. In the entry for 1128, he noted 'two black spheres against the sun' followed by what we now know to be auroras over Hereford. Illustrations in works like this were extremely rare, but were integral to how John wove expert astronomical knowledge into his history. He saw these sunspots as portents. They appear in just one of only a handful of illustrations in the manuscript, where the others portray King Henry I's visions of death and his resulting penitence over his oppressive government. John presented the sunspots as heralding these visions and therefore worthy of illustration.

These must have been incredibly large spots to be visible to the naked eye. Contemporary Korean astronomers recorded red light in the night sky over Songdo in the *Koryo-sa*, their official chronicle, five days later. This auroral display was probably caused by the same spots, representing a unique cross-reference of such early observations.

Previous page
Detail of James Nasmyth's 1860 painting showing the size of a sunspot in relation to the Earth (see p. 45).

in Christi brachio regnum Anglie sustineat. Et ut huiusmodi consilium stabile permaneat, ab omnibus fit iuramentum. Primo omnium iurant archiepiscopi, sicque per ordinem episcopi, Rogero Saresbiriensi presule diiudica-
tore omnium existente. Post episcopos more ecclesiastico licet omne iuramentum a Domino sit prohibitum
abbatibus foret iurandum. At quoniam a seductis in nutu sepius vilipenditur cucullatus, sequitur
ordo preposterus non necessario sed puelle commutatus. Iurat rex Scottorum David; iu-
rans etiam Anglorum regina filie regis que in presentiarum erat iuratur prerogative eo pacto assen-
sum dedit, ut si rex invictus sine herede careret herede. Si vero non invictusque superstes qui foret
regnum hereditaret. Robertum regis filium comitem Gloecestrensem ad sinistrum pedem regis
sedentem diiudicator allocutus: Surge inquit surge et pro regio velle iuramentum effice.
Ille: Maior me inquit natu prius id agat Stephanus comes Boloniensis, hic ad dextrum pedem
regis sedens. Quod et factum. Iurant per modum omnes comites, barones, vicecomites et quique nobiliores
milites. Quo facto diiudicator exclamat: Abbates procedant, iuramentum faciant.
Tunc exurgens vir reverendus abbas de sancto Eadmundo nomine Anselmus respondit pro omnibus
graviter ferens in se et in socios preposterum ordinem iuramenta factum. Ecce inquit o rex oppro-
bria exprobrantium tuo ordini ceciderunt super nos. En contra ius ecclesiasticum vilipensis abba-
tum personis, laicales personas etiam nobis homagio subactos tuo iuramento preposuisti. Ad hec rex:
Et quod iam factum stet sic ut cernitis. Adiurari cessate. Mora nulla sit. Appropiate,
iuratoreque vos, ut nos iuravimus omnes. Abbates iurant regem placent qua curant.
Finito concilio discessum agentes quique redierunt in sua. Sed pro dolor. Ecce videmus iuramentum
versum in periurium. Terrentianum dictum: Obsequium amicos, veritas odium parit.
Verum licet hoc verum sit, novit Deus Christe ei et utriusque spe, si non vereret regie maiestati caput solius
condempnari, asserere iuratores omnes periurio notari. Deus autem rerum cuius oculis nuda et aperta sunt
omnia, ut bene scit et in nullis Deus melius vult in misericordia et miserationibus, ut optime novit
cuncta disponat. Post modicum tempus rex Anglorum mare transit.
Anno regni .iii. Leodegarii Ro-manorum Imperatoris. Regis
Anglorum Heinrici .xxviii. Olimpiadis .cccc.lxx.
Anno .ii. Indictione .vii. luna .xxv. existente
VI. id. decembris. Sabbato a mane usque ad vesperam
apparuerunt quasi due nigre pile infra solis
orbitam. Una in supe-riori parte et erat
maior [illegible] al[illegible] [illegible] in
fer[illegible] [illegible] et [illegible] minor
erat [illegible] e [illegible] utraque [illegible] e
dir[illegible] a con[illegible] tra alter[illegible] a [illegible]
ad huius [illegible] figura [illegible].
Urbanus Lamorgatensis seu Landatensis episcopus qui de quarun-
dam rerum querelis quas anno preterito in generali concilio super
Bernardum episcopum de sancto David promoverat non iusta erga se agi per-
senserat, emensa festivitate purificationis sancte Marie mare transiit, Romam iuit, apostolico
pape causam itineris certa attestatione suorum intimavit. Cuius idem apostolicus votis ac dictis
favit, regique Anglorum H. Willelmo archiepiscopo et omnibus Anglie episcopis litteras direxit, omnibus apostolica
mandans auctoritate, ut iuste gradationi illi nemo obstaret in aliquo. Vir venerand

ROSA URSINA

Printed plate in *Rosa Ursina* by Christoph Scheiner, Bracciani, 1630
350 × 245 mm (manuscript page)
Science Museum Group. Object no. F O.B. SCH SCHEINER

When astronomers turned the newly invented telescope towards the Sun at the start of the 17th century, magnification allowed them to view the surface in detail for the first time. They were surprised to see dark spots that seemed to move across its face. The Jesuit astronomer Christoph Scheiner, unwilling to let go of the idea of a perfect, incorruptible Sun, argued originally that these were the shadows of small planets. A bad-tempered, published exchange with the Pisan astronomer Galileo Galilei, however, eventually persuaded him that they were marks on its surface or atmosphere. Both Scheiner and Galileo used helioscopes to project the Sun's image onto paper, allowing them to observe and draw sunspots and their movements with minute accuracy.

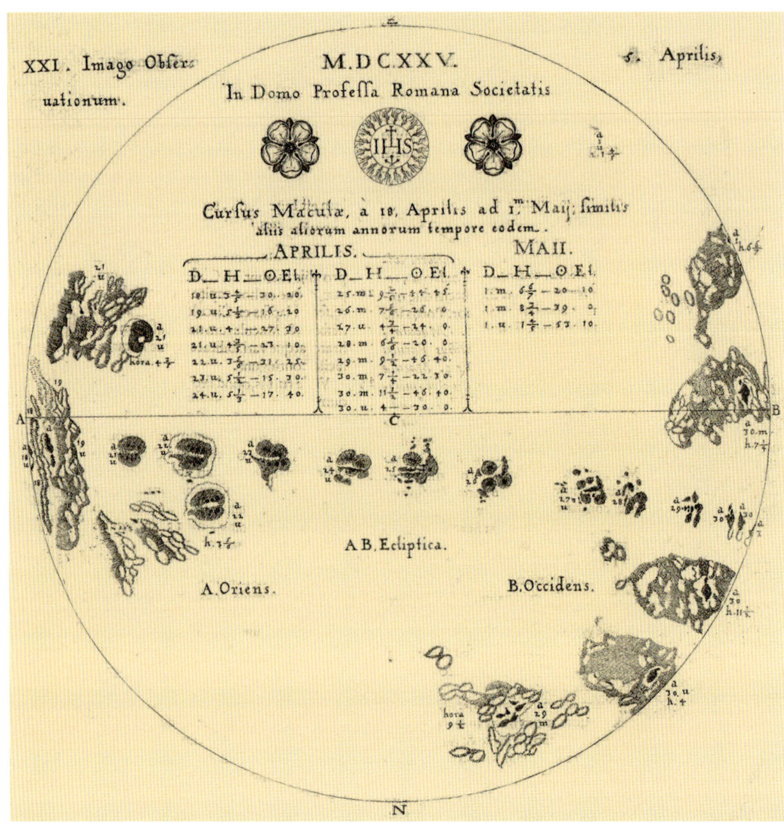

GROUP OF SPOTS ON THE SUN

Observed and drawn on paper in pencil with ink and wash by James Nasmyth, Penshurst, 13 September 1880
365 × 320 mm
Science Museum Group. Object no. 1933-580

James Nasmyth made his fortune from his Manchester factories and inventing the steam hammer. Taking early retirement in 1856, he moved to Kent to focus on his passion for astronomy, where he designed and built his own telescopes. He was particularly interested in the surfaces of the Moon and the Sun, and in how different types of image could be used to record them, from pencil drawings to photography. Here he uses pencil and paper to capture spots on the Sun's surface, made bright by the black wash of the sky. The spots are shown by lively scribbles, combined with a careful scale and a mark for the centre of the Sun's disc.

For another image by Nasmyth, see p. 45.

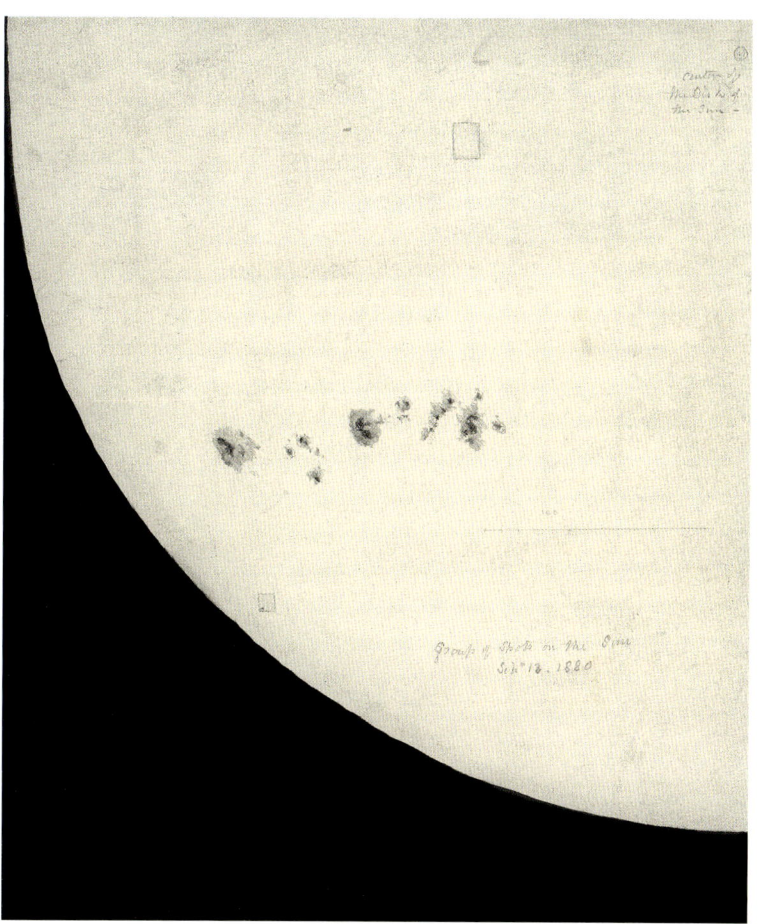

SUNSPOT COMPARED TO THE SIZE OF THE EARTH

Painted in oil, lampblack and distemper on paper by James Nasmyth, Kent, 1860
420 × 455 mm
Science Museum Group. Object no. 1899-61

James Nasmyth was the first astronomer to observe that the Sun had an uneven surface – what we now call granulation – and attempt to represent it. He wrote to the Manchester Philosophical Society in 1861 and included 'a rough but faithful drawing' very similar to this painting. He argued that the surface around sunspots was composed of '"willow-leaf" shaped filaments' that also formed 'bridges' over spots.* He accompanied the drawing with an image of these overlapping 'leaves', which appears to be a photograph of a three-dimensional model. Here, in paint, he used defined brushstrokes to represent each 'leaf', and lampblack to create the dark, cooler centre of the spot: the umbra and penumbra.

For another image by Nasmyth, see p. 43.

* 'On the structure of the luminous envelope of the sun. By James Nasmyth Esq., C E In a letter to Joseph Sidebottom, Esq.', *The Memoirs of the Literary and Philosophical Society of Manchester*, third series, vol. 1 (1862), pp. 407 and 408.

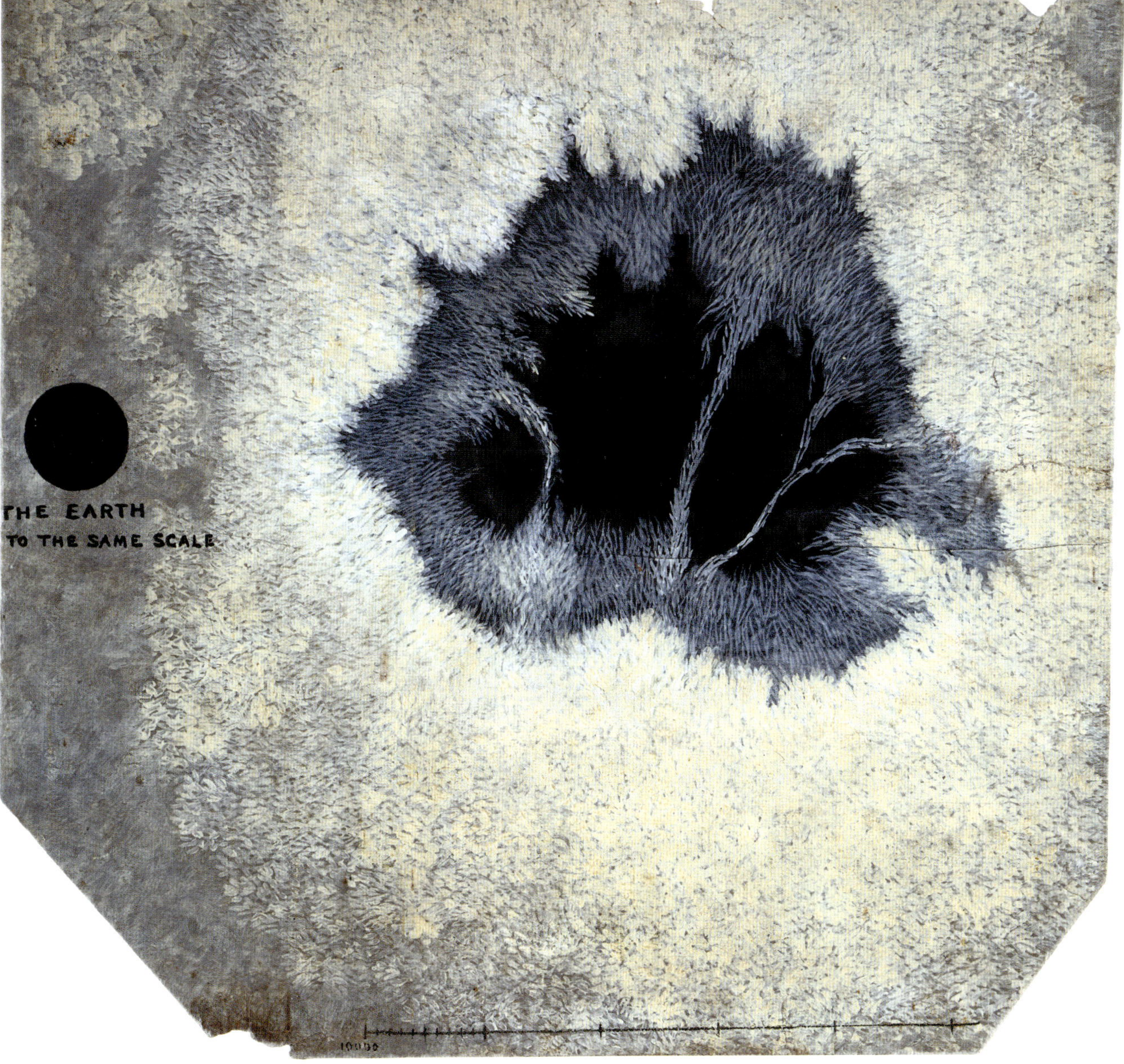
THE EARTH
TO THE SAME SCALE
10000

GROUP OF SUN SPOTS AND VEILED SPOTS

Chromolithograph by Étienne Léopold Trouvelot, published by Charles Scribner's Sons, New York, 1881
660 × 916 mm
Science Museum Group. Object no. 2000-306/1

Étienne Léopold Trouvelot was invited to join the staff of the Harvard College Observatory in 1872 by the director, Joseph Winlock. Trouvelot had sent Winlock some of his beautiful and detailed natural history illustrations, as well as publishing sun observations in various newspapers. He was tasked with producing engravings for the observatory's *Annals* and making daily observations of sunspots. In 1881, he created a series of 15 large images for a general audience, choosing the process of chromolithography, which gave him a high level of colour control. This spectacular image shows a pair of sunspots, observed on 17 June 1875, in soft browns and greys. He has given the Sun's surface a mottled look, and the spots' filaments an organic quality, reminiscent of the wings of insects from his earlier work.

Prints from Trouvelot's *Annals* also appear on pp. 66, 68, 69 and 107.

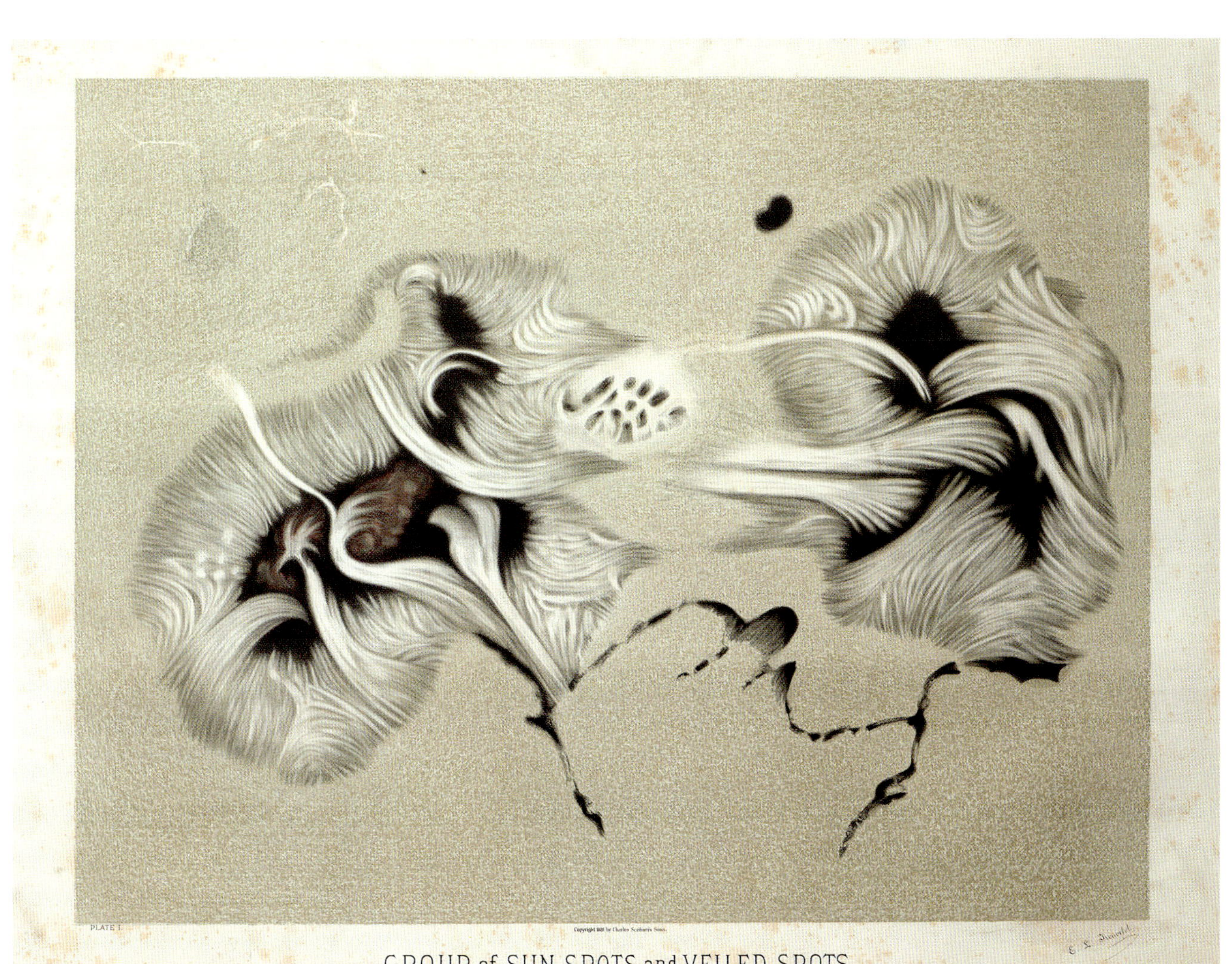

PLATE I.

E. L. Trouvelot

GROUP of SUN SPOTS and VEILED SPOTS.

Observed on June 17th 1875 at 7 h. 30 m. A.M.

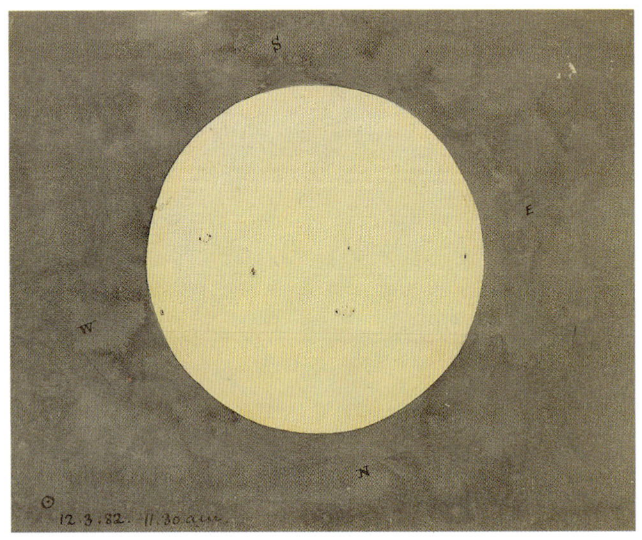

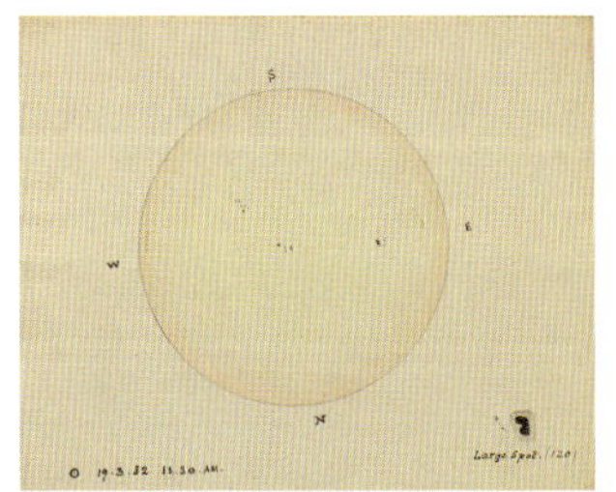

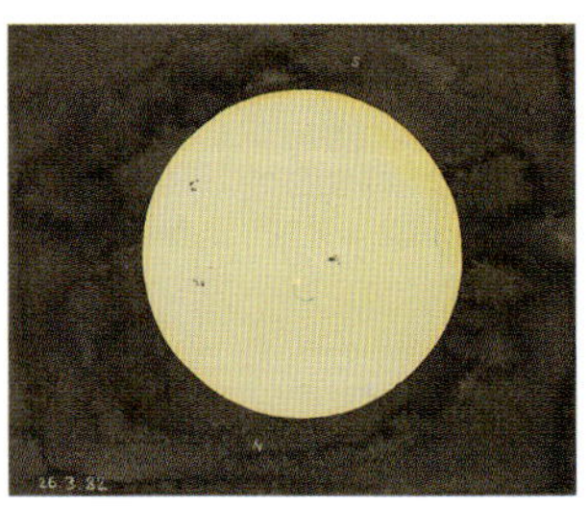

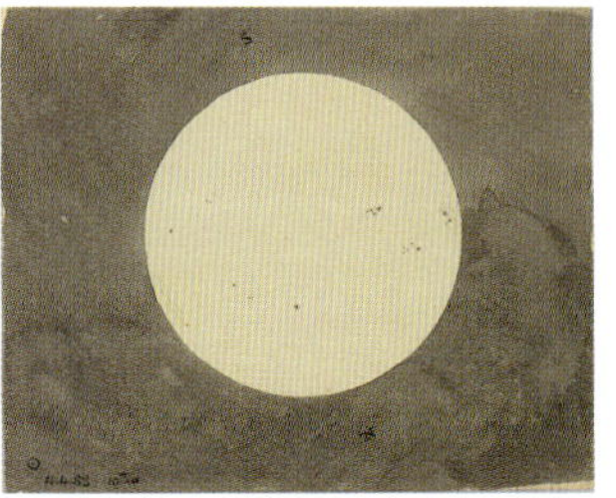

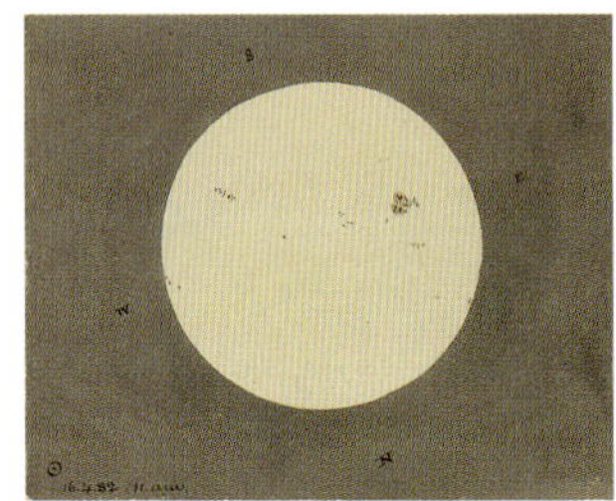

OBSERVATIONS OF THE SUN

Drawings in pencil, ink and watercolour by John Willis, possibly Folkestone, 1882
Each 145 × 160 mm
Science Museum Group. Object nos 2001-769/1 to 25

Sunspots appealed to many amateur astronomers as their changing shapes and movements could be tracked relatively easily across the Sun's disc. John Willis provided a compelling example of the obsession this could become, producing 229 drawings of the Sun between 1882 and 1905. Unused versions show that he prepared the projections in advance, for speed of observing, with black backgrounds and blank discs for the Sun, and with north at the bottom to accommodate his refractor telescope. This set comes from March to June 1882 when he observed the Sun at least once a week, including a partial eclipse on 17 May.

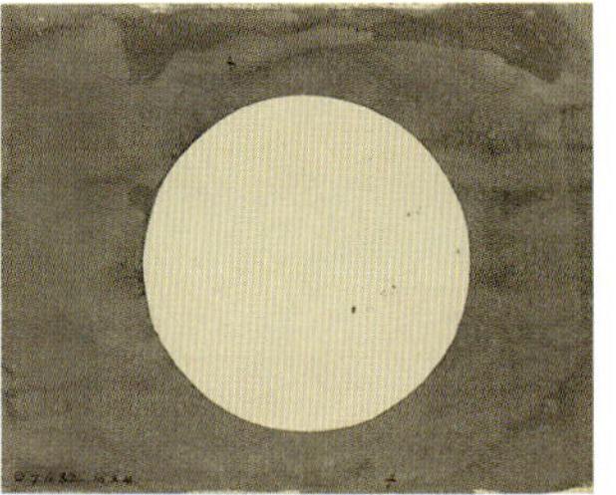

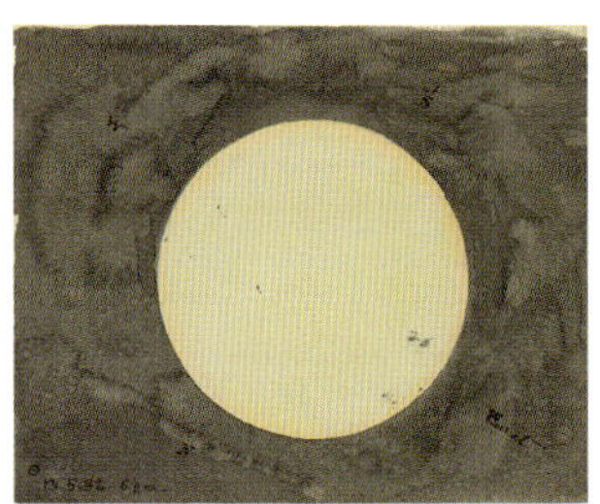

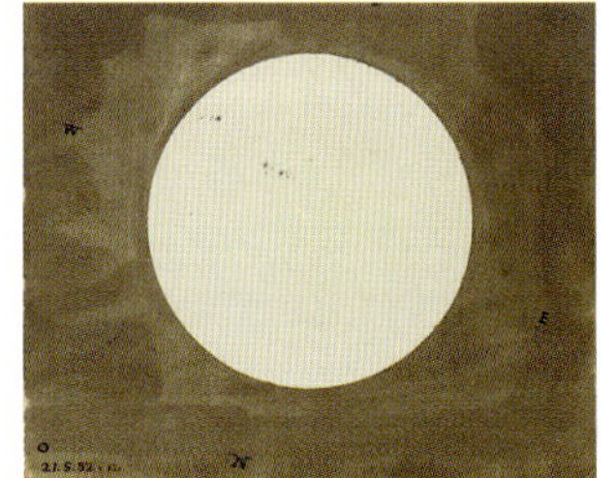

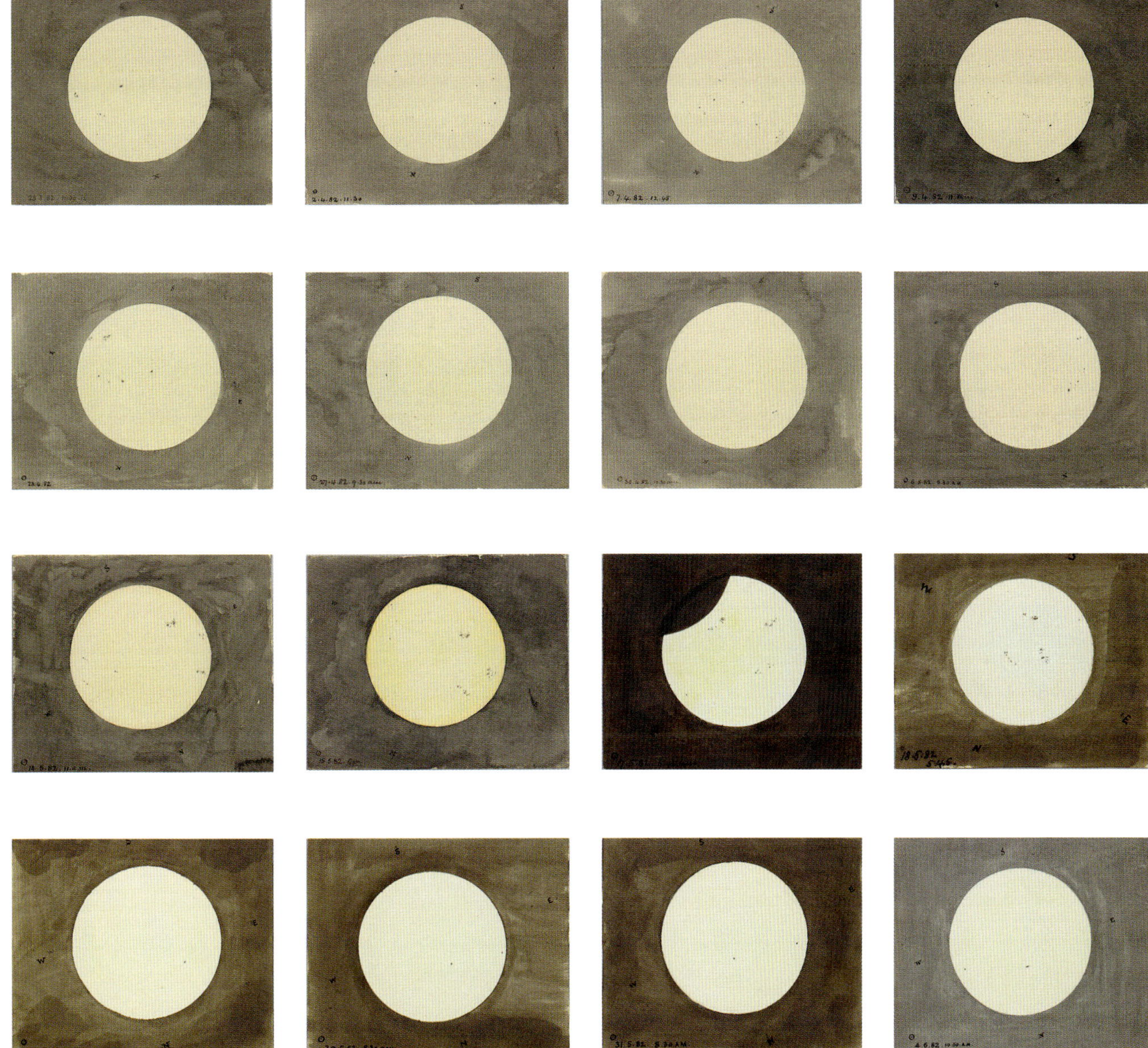

THE SUN AND SOLAR PHENOMENA

Drawn and engraved by John Emslie, published by James Reynolds, London, 1851
285 × 228 mm
Science Museum Group. Object no. 1987-889/5

Interest in solar phenomena was part of the wider public enthusiasm for astronomy in the 19th century. James Reynolds and John Emslie included a Sun print in their *Astronomical Diagrams*, which showed its mottled surface and the changing appearance of a spot. Here, the Sun appears to have layers, matching the latest opinion of Dr John Herschel

> ... that the lucid matter of the sun exists in the form of clouds of which there are two different strata, the lower stratum being less bright than the upper. The removal or opening of these clouds, he supposes, exhibits the opaque globe of the sun to our view, and hence those dark spots.'*

For another print from the *Astronomical Diagrams*, see p. 30.

* *Companion to Reynolds's Series of Astronomical and Geographical Diagrams*, 1851, verso.

THE SUN AND SOLAR PHENOMENA.

PHENOMENA OF DAY AND NIGHT.

Caused by the diurnal rotation of the Earth upon its Axis.

Modern Astronomers are of opinion that the Sun is an opaque body surrounded by luminous matter, and that the spots observed are openings in this luminous covering, showing the body of the Sun. The spots are of various sizes, some being more than 40,000 miles in diameter.

SUMMER AND WINTER RAYS.

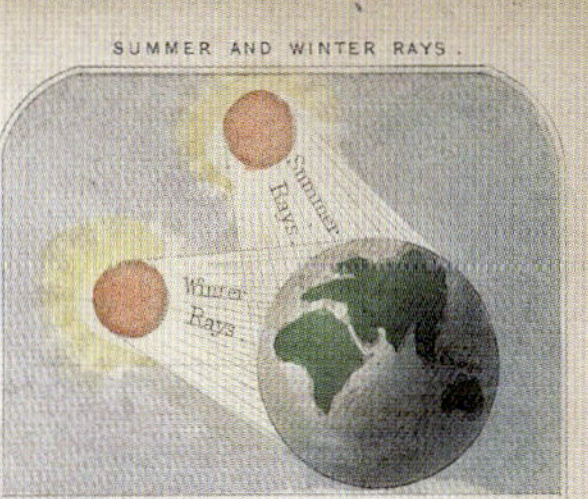

The Sun attains a greater altitude in Summer than in Winter.

COMPARATIVE SIZE OF THE SUN

as seen from

Mercury.

Venus.

Earth.

Mars.

Vesta.

Juno.

Ceres.

Pallas.

Jupiter.

Saturn.

Uranus.

Neptune.

AS SEEN FROM THE VARIOUS PLANETS.

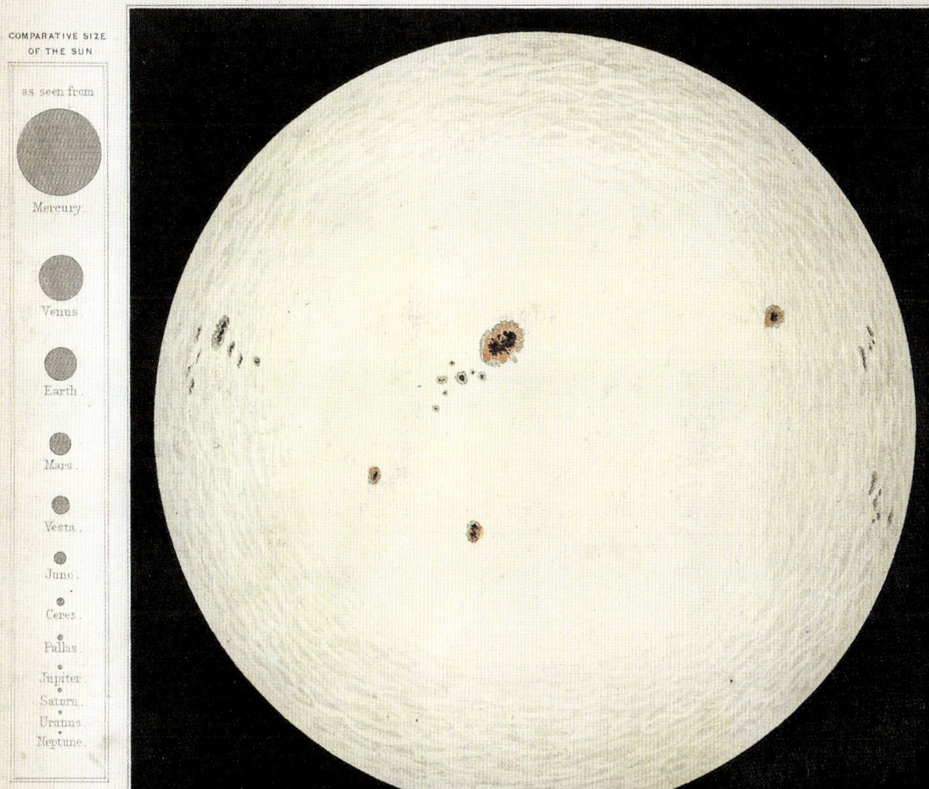

APPEARANCE OF A SPOT UPON THE SUN

AS SEEN PASSING OFF ITS DISC.

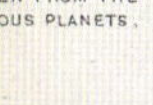

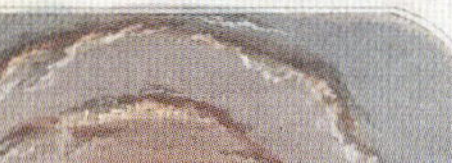

TRANSIT OF MERCURY.

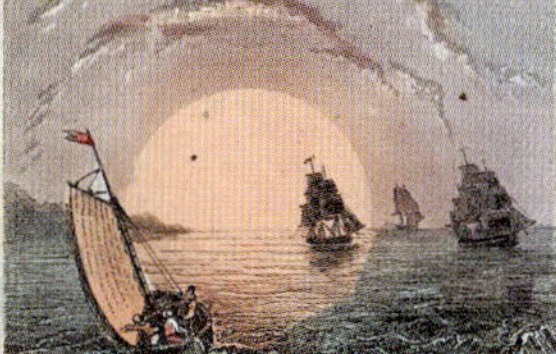

A Transit occurs when Mercury is between the Earth and the Sun in a straight line joining the centres of the two bodies. It is a somewhat rare phenomenon.

THE SUN AT MIDNIGHT AT THE NORTH CAPE OF EUROPE.

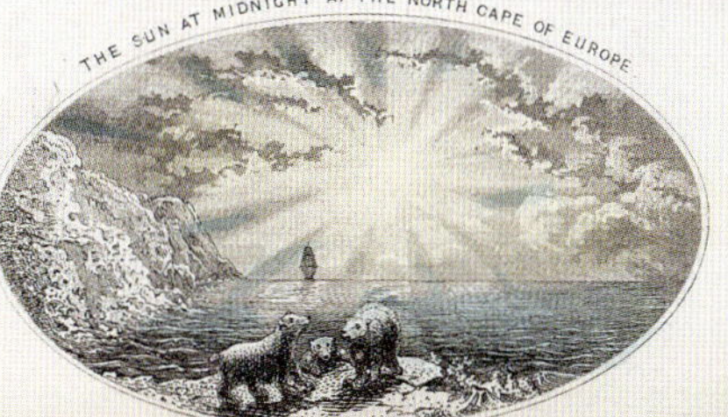

Drawn & Engraved by John Emslie

London. Published by J. Reynolds, 174 Strand.

ANNULAR ECLIPSE.

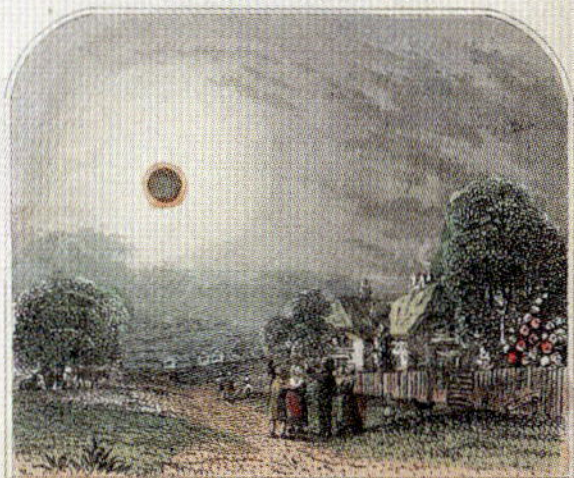

Caused by the Moon, when directly between the Earth and the Sun, passing across the centre of the Sun's disc which then presents the annular form.

DAILY PHOTOGRAPHS OF THE SUN

Eight photographs taken by Elizabeth Beckley using the Kew Photoheliograph, Kew Observatory, London, 21–29 September 1870
Each 125 × 110 mm
Science Museum Group. Object nos 1982-684/6, 10, 13, 15, 18, 26, 32, 38

This series of eight photographs was taken over a period of nine days using the Kew Photoheliograph: the first instrument designed specifically to photograph the Sun. The images show the progress of sunspots from left to right across the solar surface as the Sun rotates. The dates and times are given at the top of each image.

Money was tight at Kew and so Elizabeth Beckley, the teenage daughter of the observatory's mechanical engineer, was employed cheaply to take daily photographs of the Sun. She worked at Kew for over ten years and was one of the first female members of staff of an astronomical observatory. The Kew astronomer Warren De La Rue praised Beckley's ability to capture photographs 'even on very cloudy days, when it would be imagined that it was almost impossible'.* Detailed analysis of photographs and data by Beckley and other 'lady computers' around the world would go on to become fundamental to astronomical work.

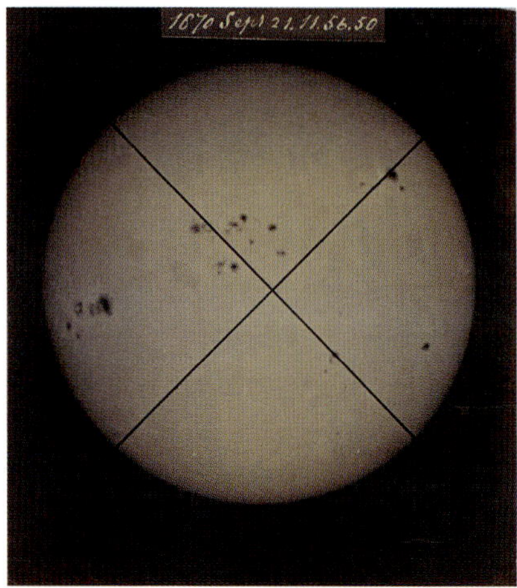

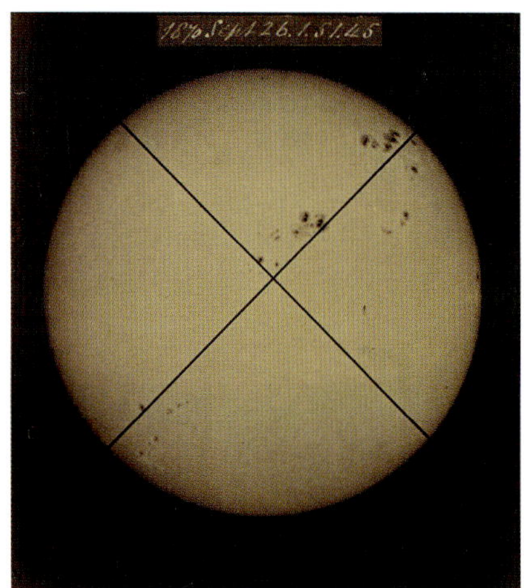

* Lee MacDonald, '"Work peculiarly fitting to a lady": Elizabeth Beckley and the early years of solar photography', *Constructing Scientific Communities* (9 March 2017), https://conscicom.org/2017/03/09/work-peculiarly-fitting-to-a-lady-elizabeth-beckley-and-the-early-years-of-solar-photography/ (accessed 15 January 2018).

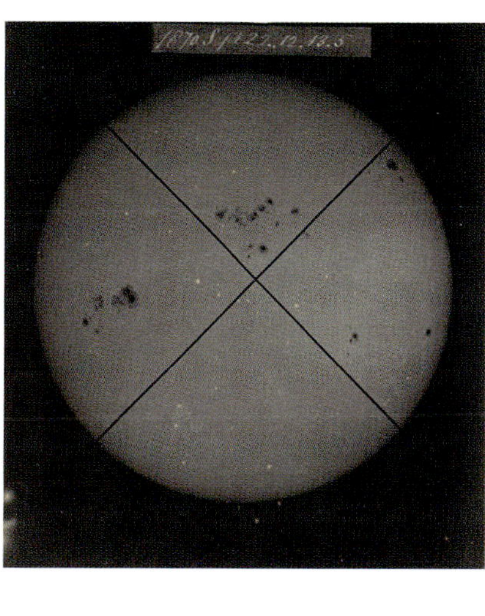
1870 Sept 22. 12.13.5

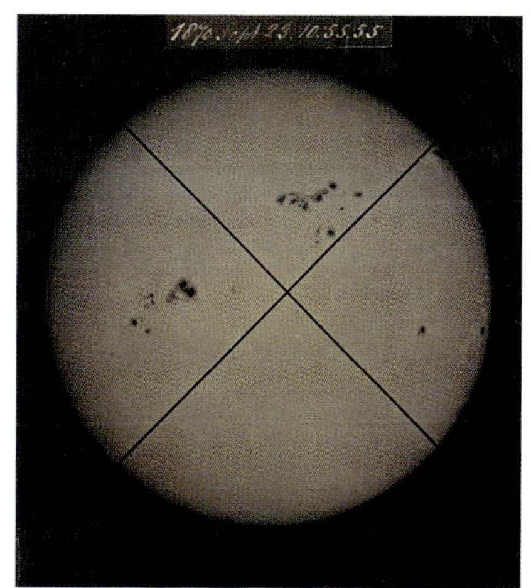
1870 Sept 23. 10.55.55

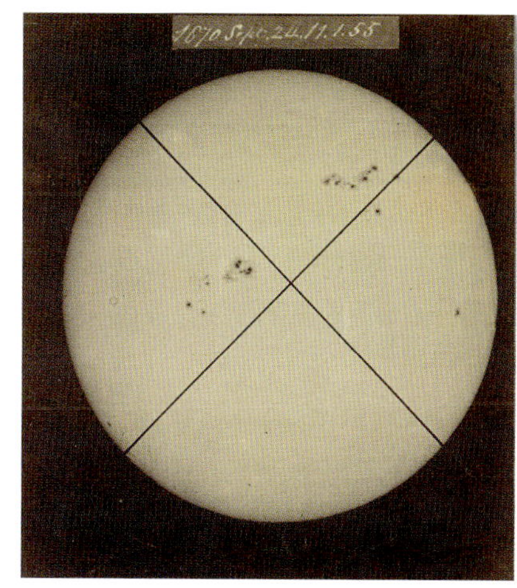
1870 Sept 24. 11.1.55

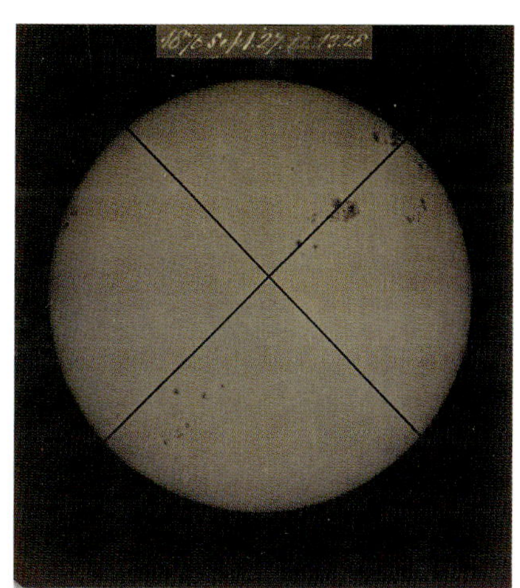
1870 Sept 27. 12.13.28

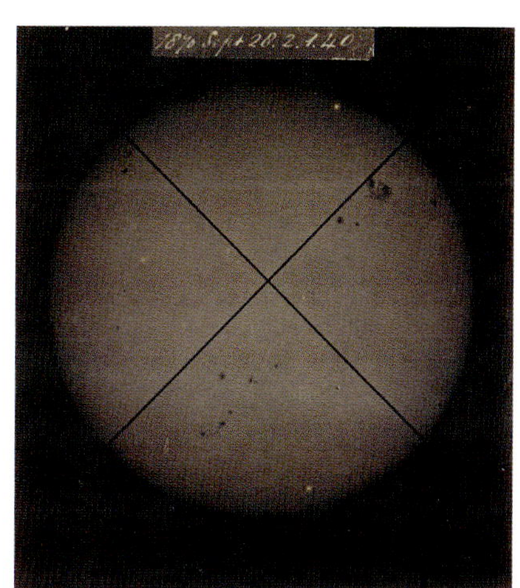
1870 Sept 28. 2.1.40

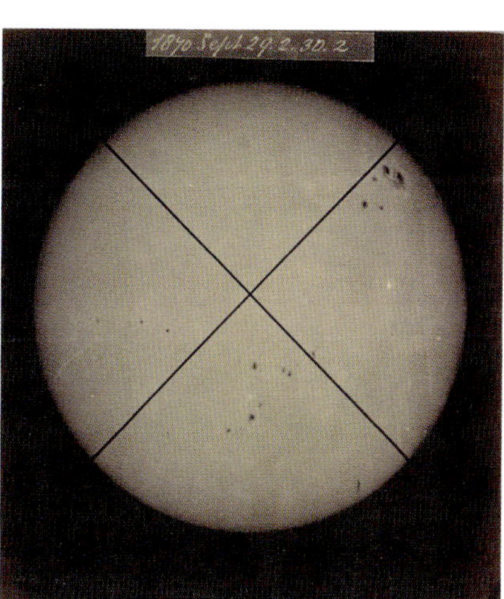
1870 Sept 29. 2.30.2

PART OF THE SUN PHOTOGRAPHED UNDER HIGH MAGNIFICATION

Photograph by Pierre Jules César Janssen, Meudon Observatory, 10 May 1878
575 × 465 mm (printed page)
Science Museum Group. Object no. 1884-25

This abstract-looking photograph, taken by the French astronomer Jules Janssen, shows a close-up of the photosphere: the visible surface of the Sun. Janssen was the first person to capture the Sun's mottled appearance with highly detailed photography. This mottling is now understood to be granulation caused by columns of superheated plasma rising from the Sun's interior.

Janssen was instrumental in establishing an astrophysical observatory at Meudon, Paris. Here, the observatory team took over 6000 photographs of the Sun between 1876 and 1903. These images formed the basis of Janssen's seminal work, the great *Atlas de photographies solaires*, published in 1904.

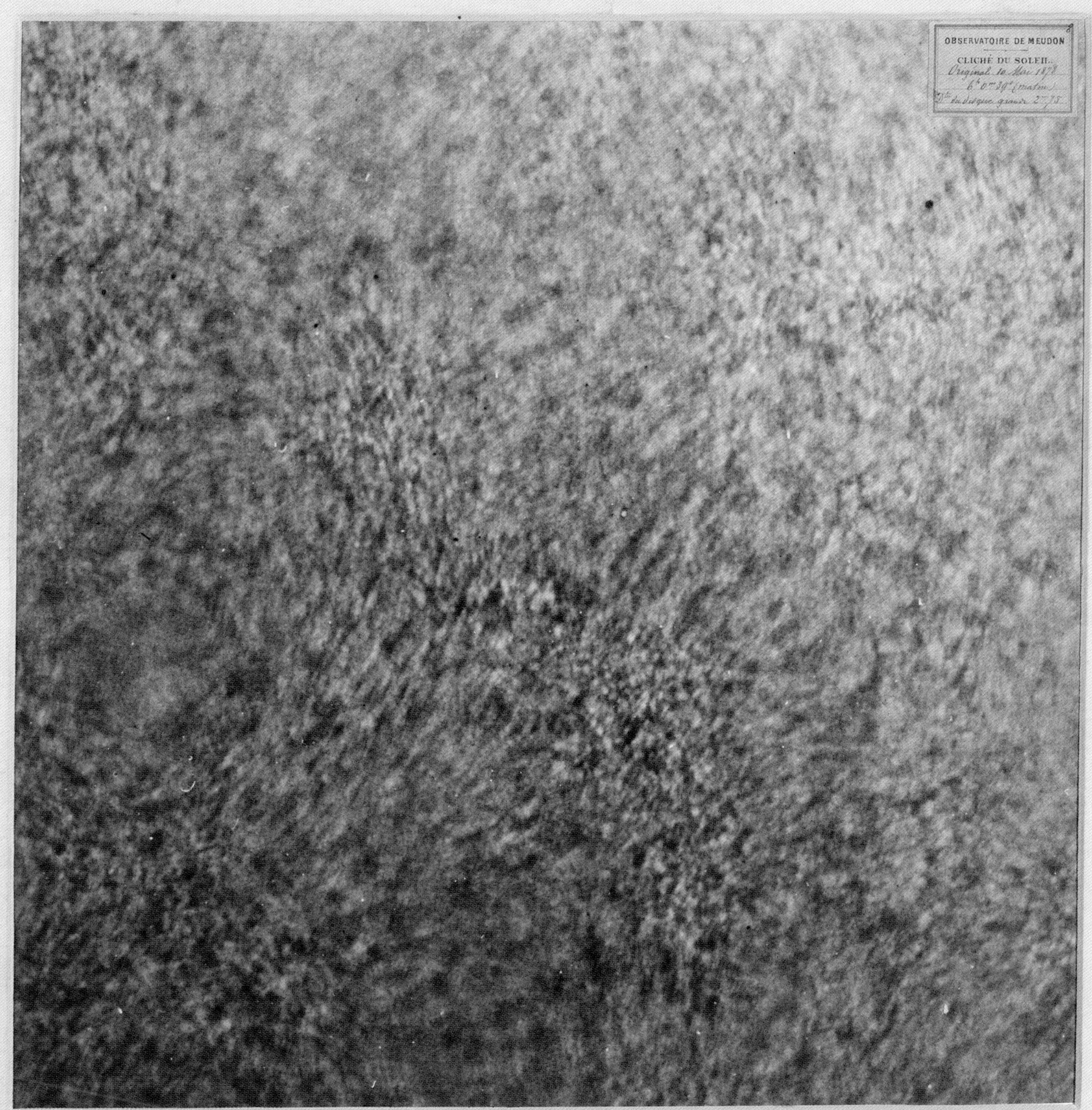

ENLARGED PHOTOGRAPH OF PART OF THE SUN'S DISC.

ORIGINAL TAKEN 10TH MAY 1878, 6 HRS. 0 MINS. 39 SECS.

PRESENTED BY M. JANSSEN, MEUDON OBSERVATORY.

BUTTERFLY DIAGRAM

Pencil and ink on graph paper by Annie and Walter Maunder, London, England, 1913
457 × 610 mm
High Altitude Observatory, National Center for Atmospheric Research, Boulder, Colorado, USA

This diagram captures the heartbeat of the Sun. Drawn by the astronomers Annie and Walter Maunder, each mark represents the position of a sunspot on the solar surface. The butterfly-like structures are the result of three solar cycles: periods of about 11 years during which the number of sunspots rises and falls.

The discovery of the butterfly diagram was the result of years of painstaking work by the astronomer-couple. Annie was first employed as a 'lady computer' at the Royal Observatory, Greenwich, where her duties included taking daily photographs of the Sun. It was there that she met Walter, and although she was forced to give up her job when they married, they continued to work together on astronomy. She became one of the most significant figures in 20th-century solar physics. She recalled that,

> we made this diagram in a week of evenings, one dictating and the other ruling these little lines that show the spots in any given latitude at any time in the sunspot period. Every half hour or so we exchanged the dictating and ruling according as voice or hand got tired.*

* Silvia Dalla and Lyndsay Fletcher, 'A pioneer of solar astronomy', *Astronomy & Geophysics*, 57, issue 5 (1 October 2016), pp. 5.21–5.23.

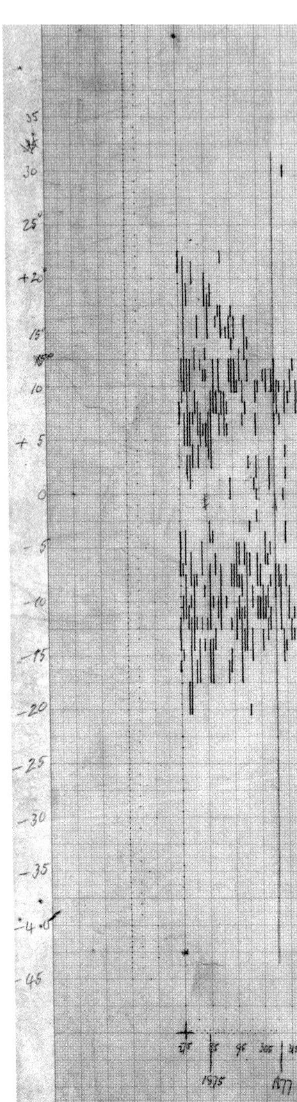

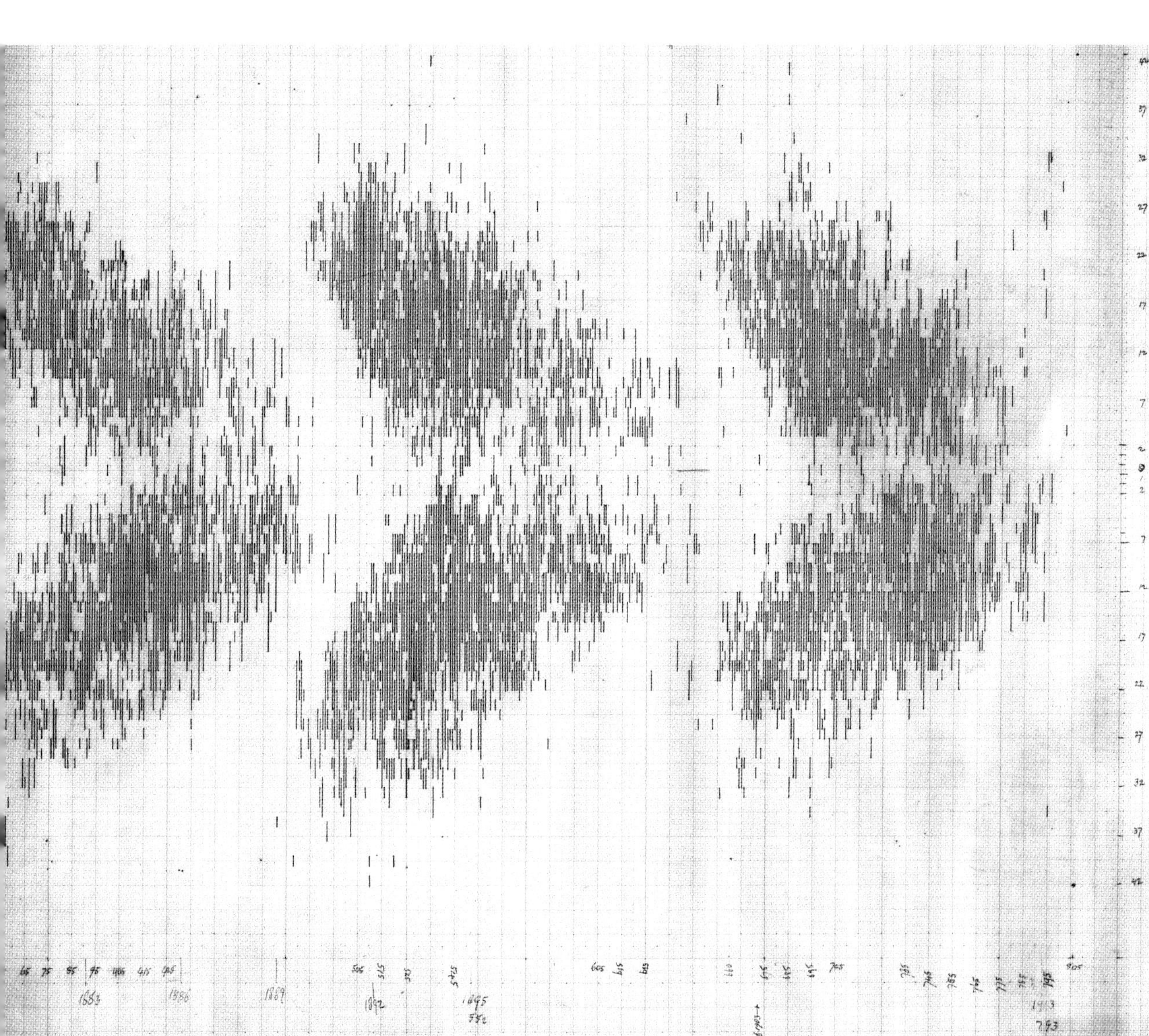

THE SUN IN ORDINARY, CALCIUM AND HYDROGEN LIGHT

Glass positive photograph and two spectroheliograms taken at the Kodaikanal Observatory, India, 10 August 1917
Each 253 × 253 mm
Science Museum Group. Object nos Evershed/E109, E110, E111

The apparently white light of the Sun is in fact a mix of colours across the spectrum, each one revealing different features of the Sun's turbulent surface and atmosphere. The image on the left, produced using white light, shows the photosphere – the visible surface of the Sun – marked in this case by a particularly large group of sunspots.

The other two images were produced using a spectroheliograph, which photographs the Sun in a single colour. They reveal the more visually active chromosphere: a brightly coloured region of the Sun's atmosphere rising 5000 km above the visible surface. These images show the chromosphere in the characteristic violet light emitted by calcium atoms, and in the red light emitted by hydrogen.

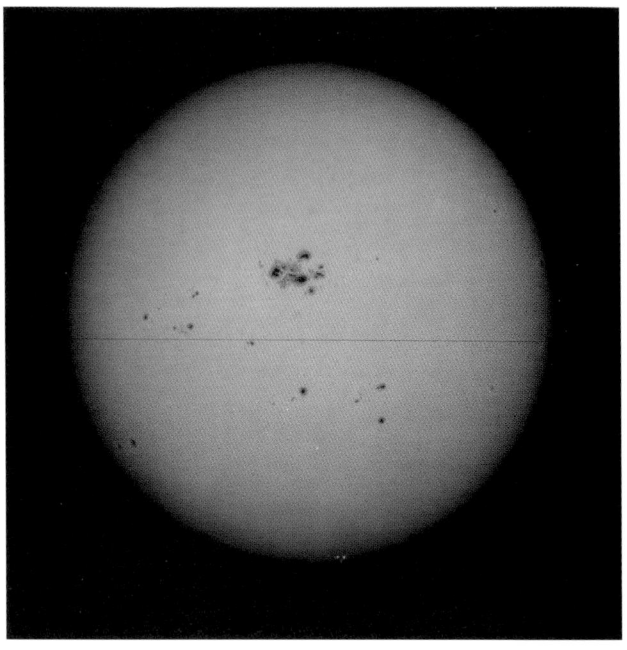

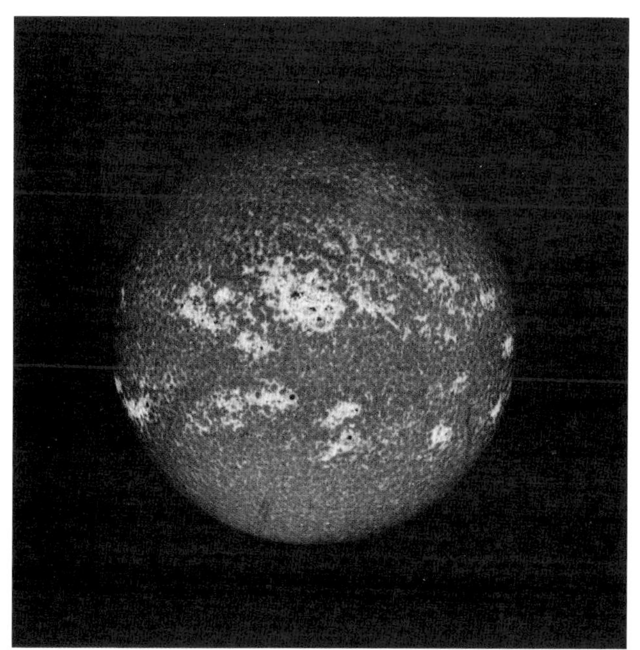

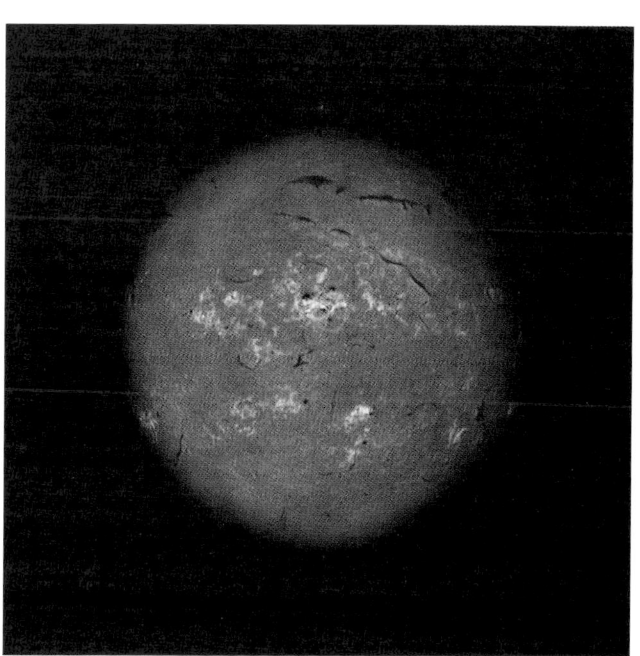

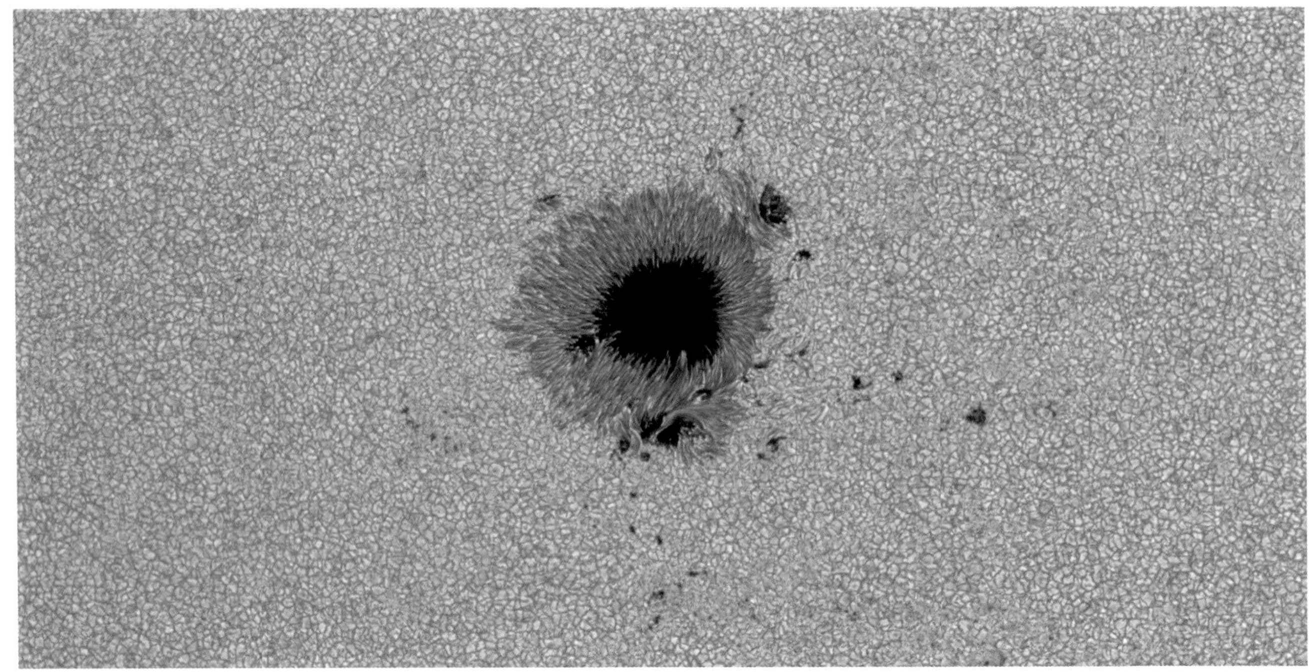

A SUNSPOT UP CLOSE

Photograph taken by the Hinode Spacecraft's Solar Optical Telescope,
10 December 2006
Digital image
Hinode SOT, Courtesy NAOJ, LMATC, JAXA, NASA, MELCO and HAO

This image reveals a sunspot in stunning detail. The dark umbra at the centre is surrounded by the striated penumbra, reminiscent of the 'willow leaf' pattern seen and painted in the 19th century by James Nasmyth (see p. 45).

Sunspots shine brilliantly, but look darker than the rest of the Sun as they are about 2000 degrees Celsius cooler. They form where intense magnetic fields break through the solar surface, preventing superheated plasma from rising from the Sun's interior. Away from the sunspot, the face of the Sun is covered in granule-like cells, each around 1000 km across, which appear where upwelling columns of plasma breach the surface.

This photograph was taken by the Hinode (meaning 'sunrise' in Japanese) Spacecraft, a mission led by the Japan Aerospace Exploration Agency, with contributions from the USA, UK and Europe.

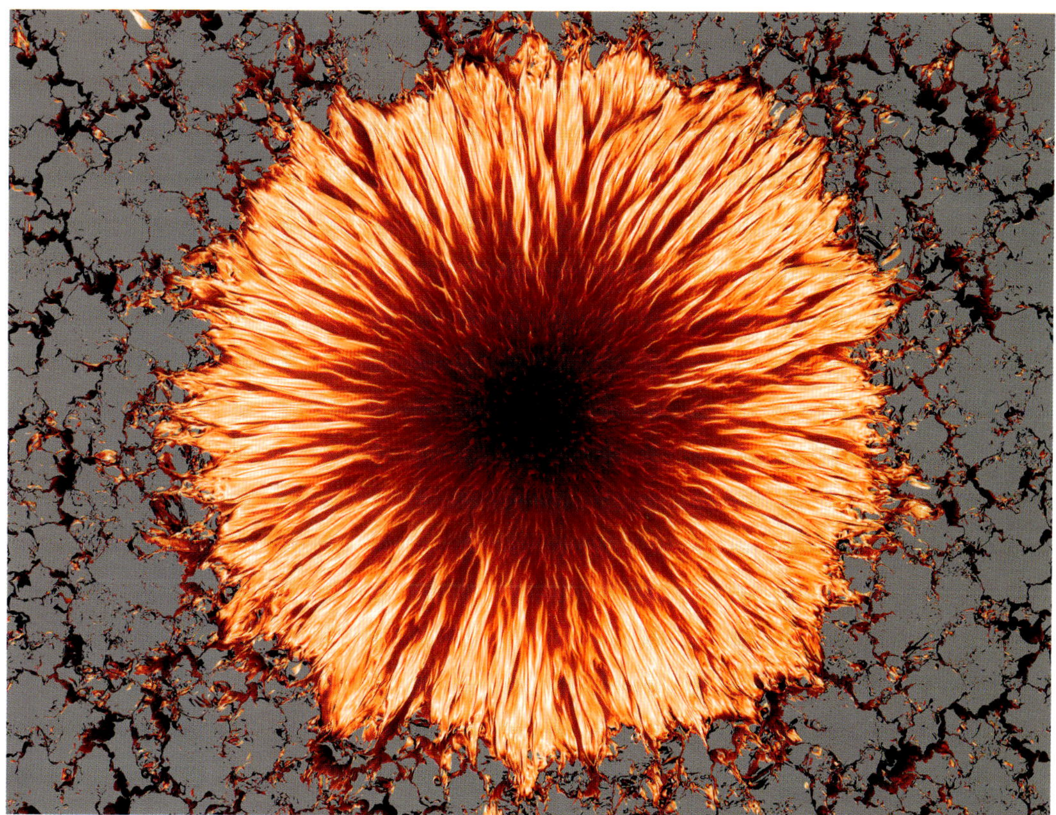

SUPERCOMPUTER SIMULATION OF A SUNSPOT

Simulation by Matthias Rempel, National Center for Atmospheric Research, Boulder, Colorado, USA, 2011
Computer simulation
Matthias Rempel, NCAR

This flower-like image was produced using the first ever three-dimensional digital simulation of a sunspot, created by scientists at the National Center for Atmospheric Research.

Rather than showing the surface of the Sun, the simulation reveals the complex magnetic field emerging from the sunspot. In the dark central region, intense magnetic flux points vertically out from the surface of the Sun, graduating to white at the edge of the spot where the field is aligned with the solar surface. Advanced simulations like these are crucial for improving scientists' understanding of our dynamic star and how it affects us on Earth.

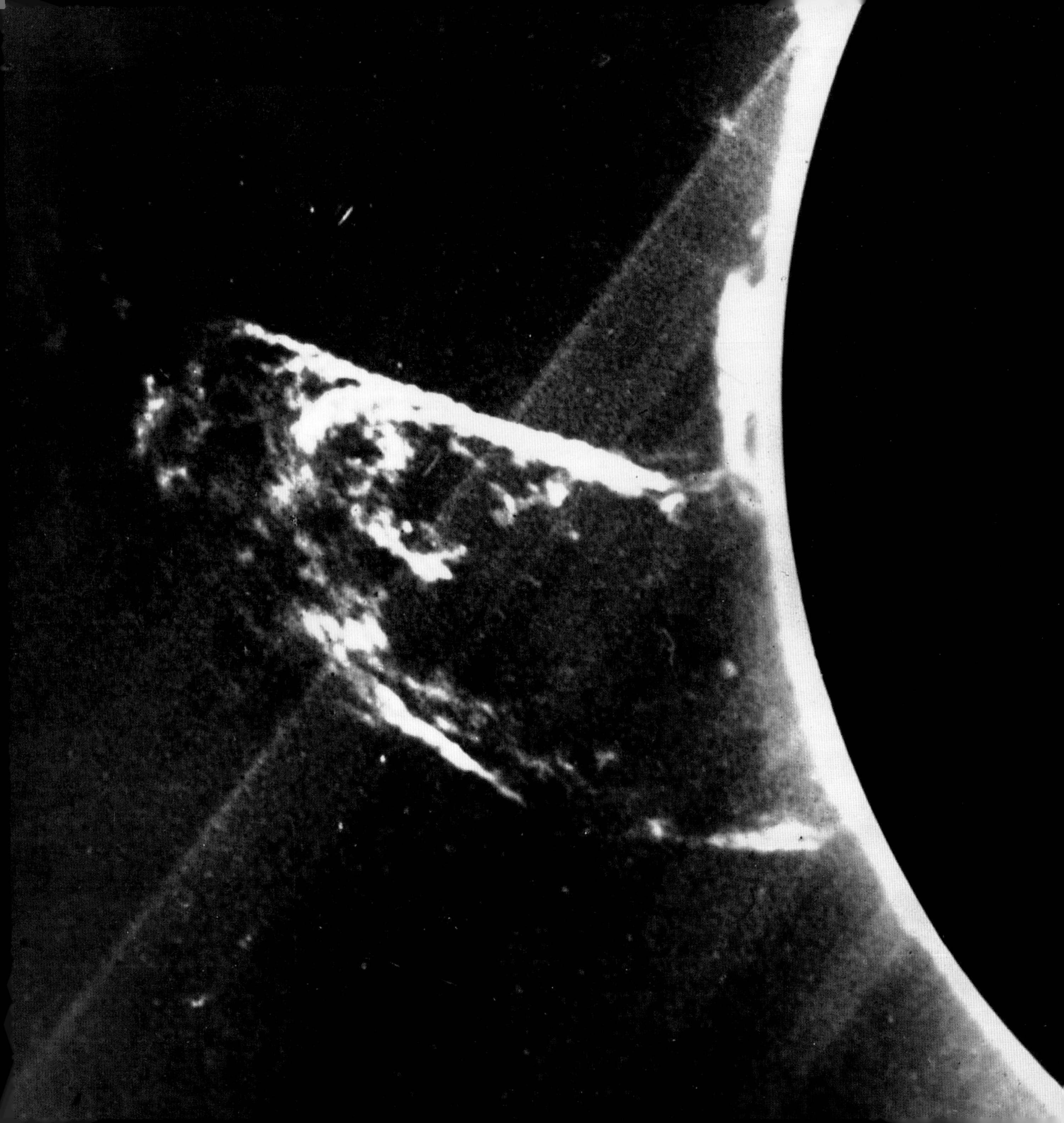

3

ERUPTIONS

On a pleasant spring day, the Sun can seem benign, gently warming our skin and bringing light to the world around us. The reality could not be more different. One hundred and fifty million kilometres away from us is a vast, raging nuclear furnace. Fountains of superhot plasma, dwarfing the Earth, erupt from the Sun's surface, towering thousands of kilometres into the blistering solar atmosphere. Loops of magnetic energy arc between sunspots, twisting and coiling until they break with unimaginable violence, releasing blinding flares of electromagnetic radiation and sending colossal clouds of plasma hurtling into space.

Our tiny planet is shielded from this constant barrage only by the protective bubble of its magnetic field. Occasionally, powerful solar storms break through to create shimmering auroral displays, in greens, blues and blood reds. Since the 19th century, astronomers have been able to observe, draw, paint and photograph these cataclysmic yet beautiful eruptions, in order to understand better our turbulent star.

SCHEMA CORPORIS SOLARIS

Engraved print published in *Mundus Subterraneus* by Athanasius Kircher, Amsterdam, 1678
380 × 420 mm (manuscript double page)
Science Museum Group. Object no. F O.B. KIR KIRCHER

Athanasius Kircher was one of the earliest writers to argue that the Sun was not a perfect, immutable body, but in fact had an active and changing surface like the Earth. While teaching at the Jesuit college in Rome he became fascinated by volcanoes, and climbed Vesuvius in 1638. He concluded that both the Earth and Sun were volcanic, that the Earth contained a subterranean network of channels carrying fire and water, and that the Sun was a mixture of solids and liquids.

This magnificent engraved image of the Sun, produced for his expensive, elaborate book, shows an active surface covered in flames and smoke-belching volcanoes. It continued to be reproduced for hundreds of years, often unattributed. Kircher's depiction foreshadows later astronomical imagery, revealing the violent surface of our Sun.

For Kircher's Sun incorporated into another print, see p. 102.

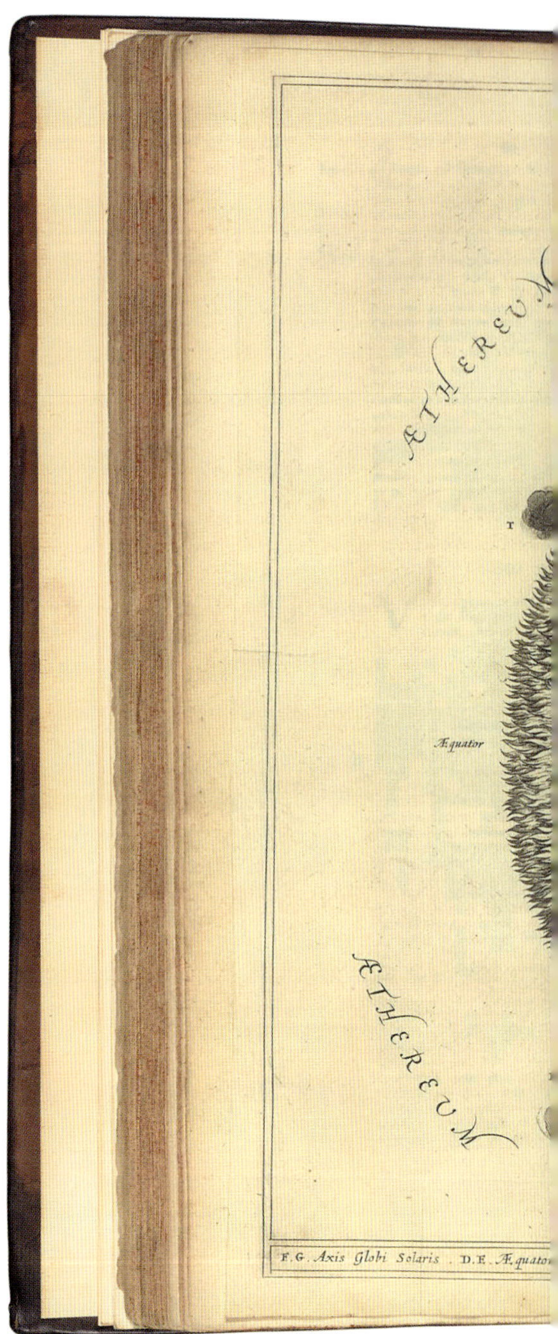

Previous page
Detail of an erupting prominence captured by John Evershed, 1916 (see p. 75).

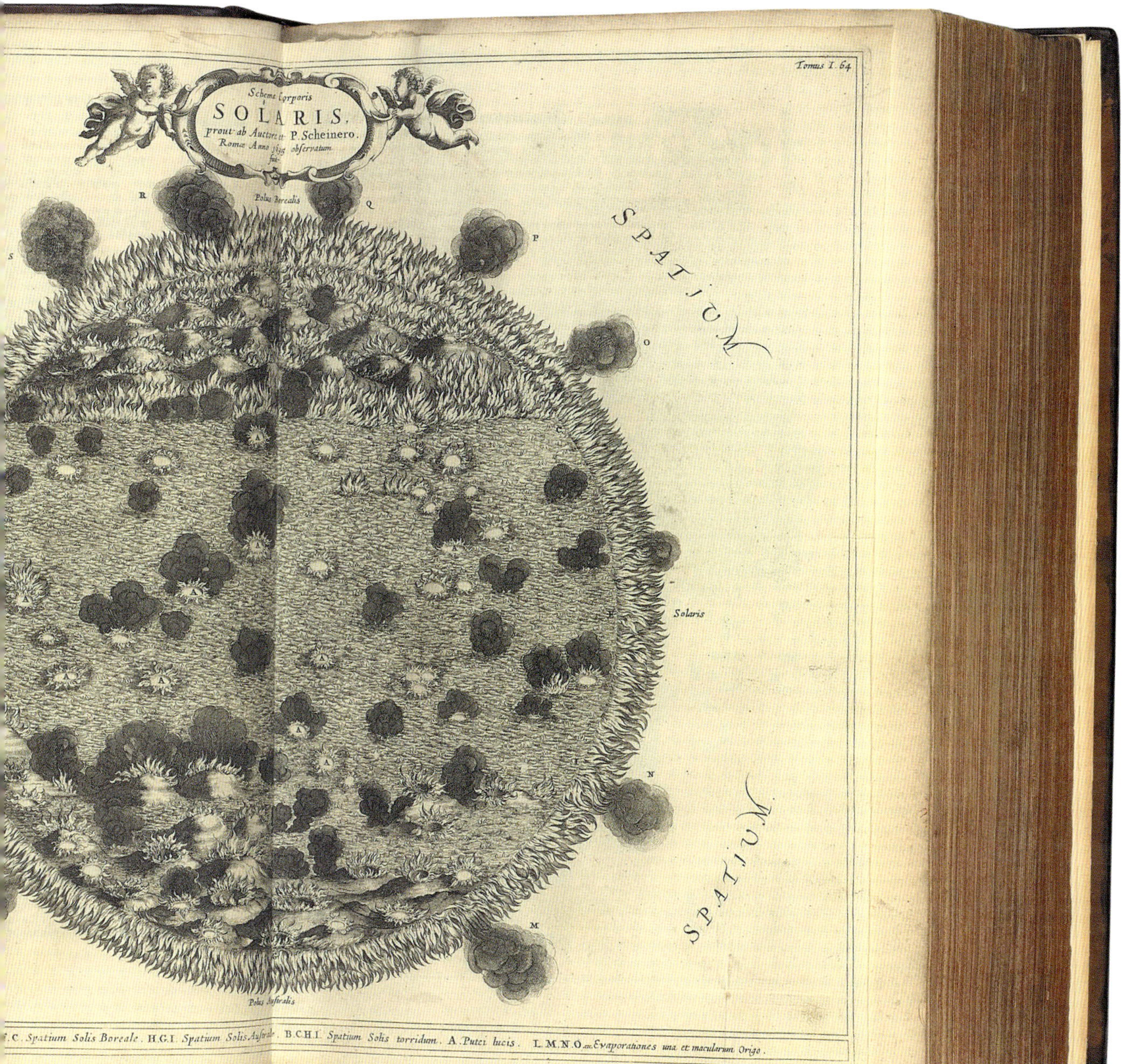
Tomus I. 64
Schema Corporis
SOLARIS,
prout ab Auctore et P. Scheinero.
Romae Anno 1635 observatum fuit.
R
Polus Borealis
Q
P
S
SPATIUM
O
Solaris
N
SPATIUM
M
Polus Australis
C. Spatium Solis Boreale. H.G.I. Spatium Solis Australe. B.C.H.I. Spatium Solis torridum. A. Putei lucis. L.M.N.O. etc. Evaporationes una et macularum Origo.

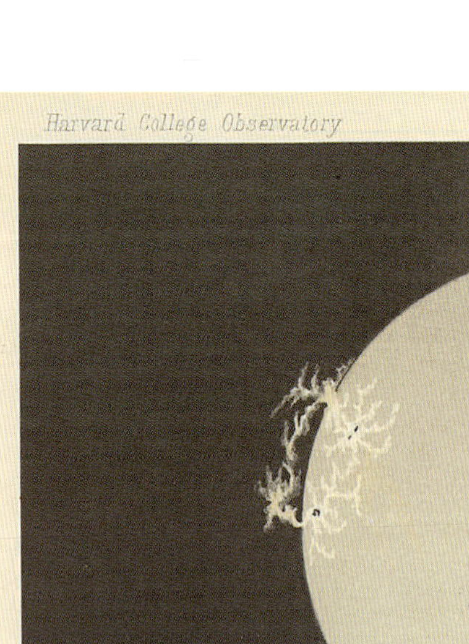

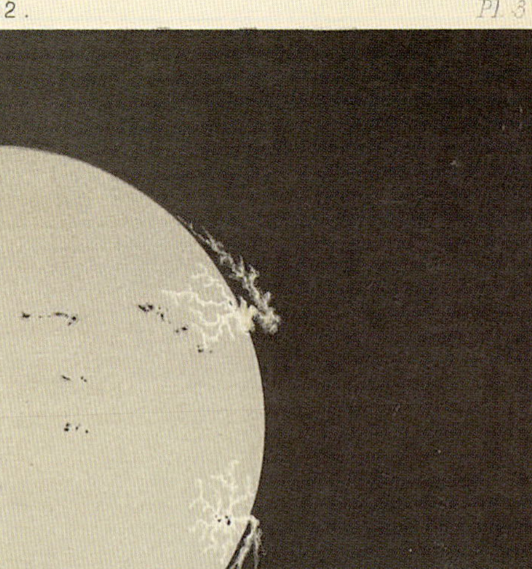

FEBRUARY 2D

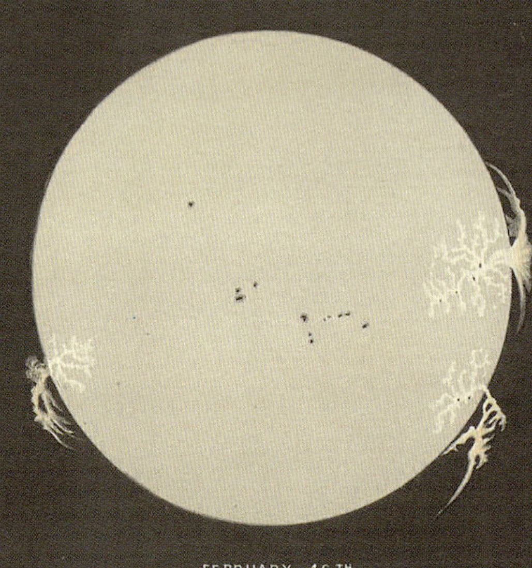

FEBRUARY 16TH

L. TROUVELOT.

THE SUN, 1872

Printed lithograph by J H Bufford after Étienne Léopold Trouvelot for the *Annals of the Harvard College Observatory*, vol. 8, Cambridge, 1876
308 × 255 mm
Science Museum Group. Object no. 1887-23/1

Étienne Léopold Trouvelot's position at the Harvard College Observatory required him to make daily observations of the Sun, and he took a particular interest in sunspots and prominences. From over 7000 drawings he worked with the director Joseph Winlock to select 35 for publication in the observatory's *Annals* in 1876. This original Plate 3 shows the Sun on 2 and 16 February 1872 with branching white prominences growing out of the spots. The plate was withdrawn and replaced with one showing red prominences silhouetted around the disc, perhaps because Trouvelot felt the colours were truer.

Prints from Trouvelot's *Annals* also appear on pp. 47, 68, 69 and 107.

SOLAR PROMINENCES, 1872

Printed lithograph by J H Bufford after Étienne Léopold Trouvelot for the *Annals of the Harvard College Observatory*, vol. 8, Cambridge, 1872
Each 360 × 280 mm
Science Museum Group. Object nos 1887-24/1 to 3

Étienne Léopold Trouvelot showed particular delight in the violence and energy of the solar prominences that he attempted to capture for the Harvard College Observatory. In the *Manual* to accompany his 1881 prints he later described them at length:

> some resemble huge clumsy masses having a few perforations on their sides; while others form a succession of arches supported by pillars ... Some resemble flames driven by the wind; others ... appear as immense fiery bundles, from which sometimes issue long and delicate columns surmounted by torch-like objects of the most fantastic pattern. Some others resemble trees, or animal forms, in a very striking manner.*

Prints from Trouvelot's *Annals* also appear on pp. 47, 66 and 107.

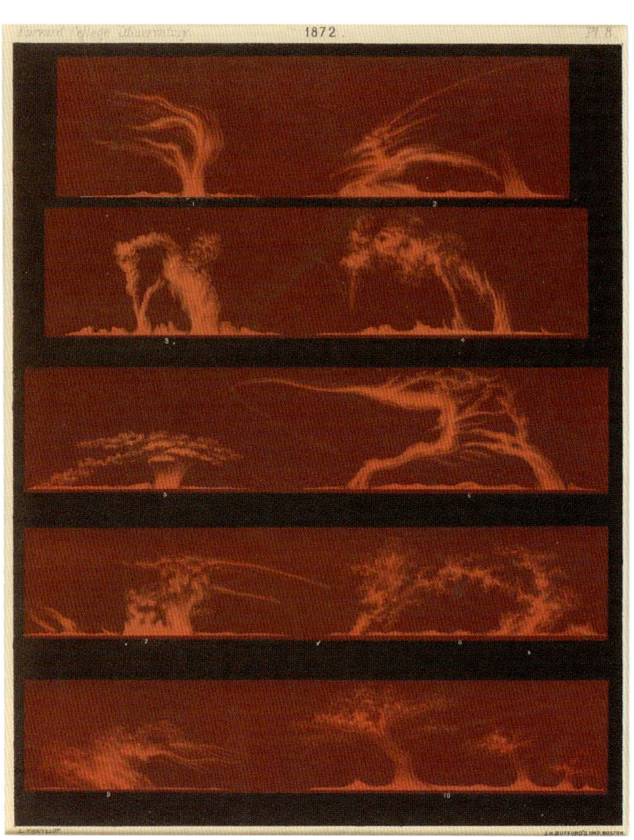

* *The Trouvelot Astronomical Drawings Manual* (New York: Charles Scribner's Sons, 1882), pp. 19–20.

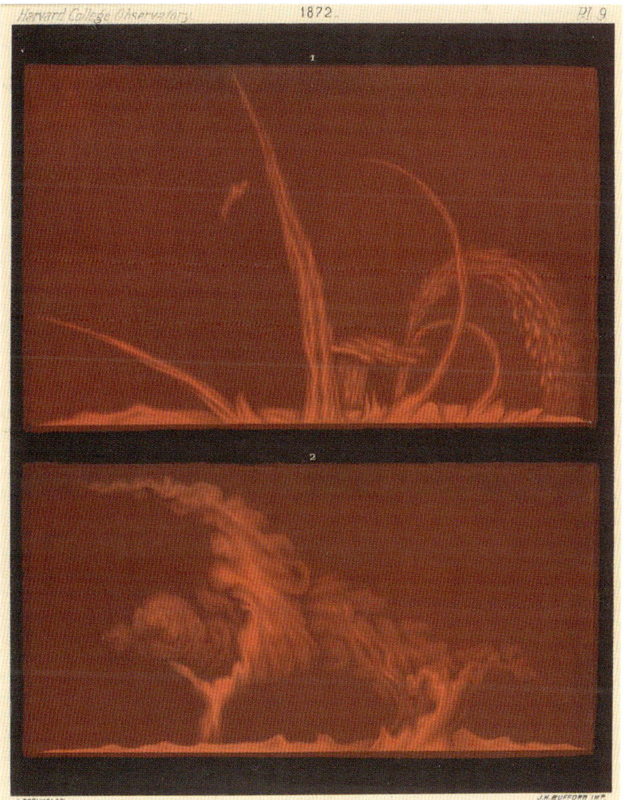
Harvard College Observatory
1872.
Pl. 9
1
2
L. TROUVELOT
J.H. BUFFORD IMP.

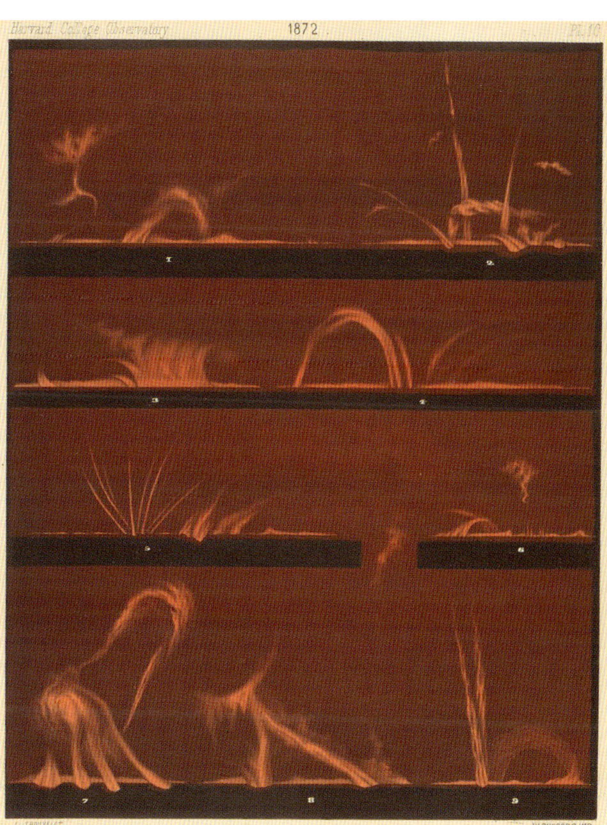
Harvard College Observatory
1872.
L. TROUVELOT
J.H. BUFFORD IMP.

THE CARRINGTON FLARE

Notebook entry by Richard Carrington, Redhill, Surrey,
1 September 1859
285 × 480 mm (notebook double page)
Royal Astronomical Society

On the morning of 1 September 1859, Richard Carrington was studying the Sun from his private observatory when he saw something extraordinary. Suddenly, a brilliant white light broke out over a particularly large sunspot, for a moment shining twice as bright as the usual solar disc. Startled, he hurriedly noted its position in this notebook alongside his sketch of the sunspot.

Carrington had witnessed a solar flare: a flash of radiation with the explosive power of up to a trillion atomic bombs. It was accompanied by an enormous eruption of material from the Sun that slammed into the Earth around 17 hours later, creating the most powerful geomagnetic storm ever recorded. Sparks flew from telegraph wires across Europe and North America as skies from the poles to the tropics erupted in spectacular auroral displays, shimmering with colours from deep red to lurid green.

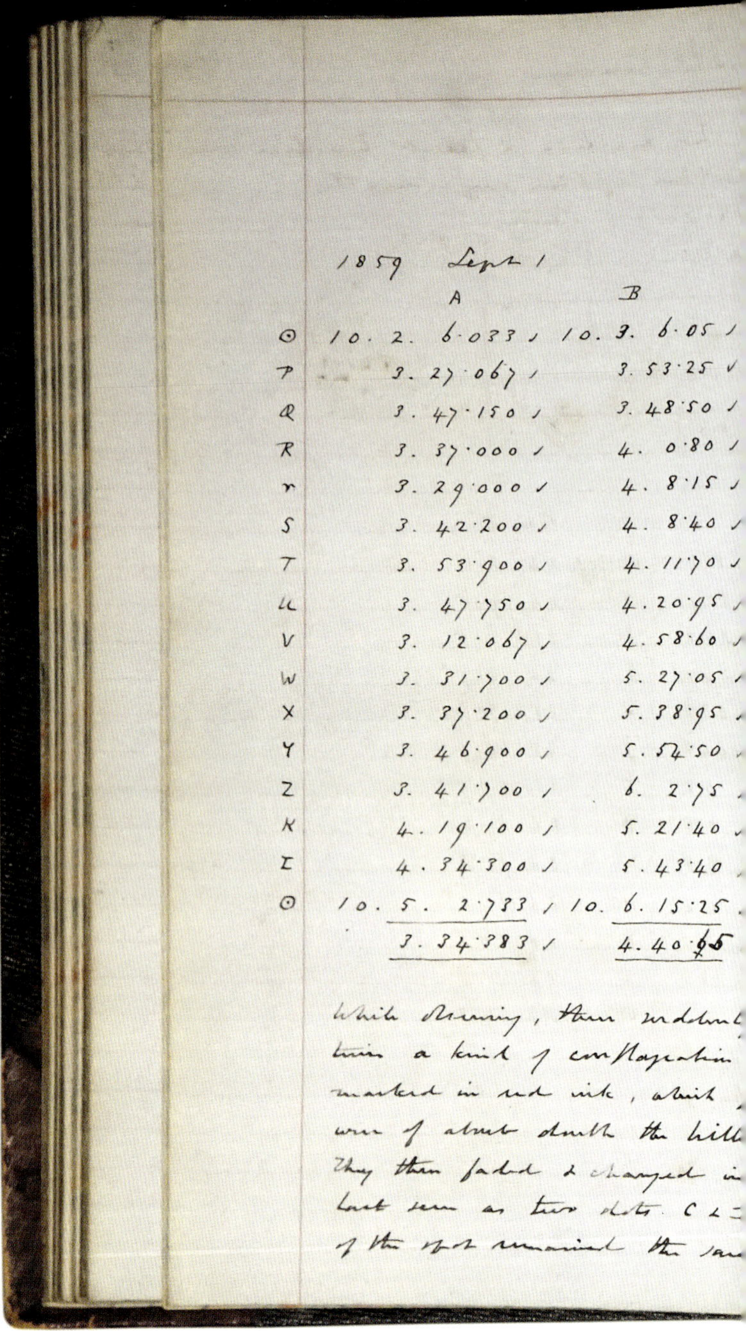

1859 Sept 1

	A	B
⊙	10. 2. 6·033	10. 3. 6·05
P	3. 27·067	3. 53·25
Q	3. 47·150	3. 48·50
R	3. 37·000	4. 0·80
r	3. 29·000	4. 8·15
S	3. 42·200	4. 8·40
T	3. 53·900	4. 11·70
u	3. 47·750	4. 20·95
V	3. 12·067	4. 58·60
W	3. 31·700	5. 27·05
X	3. 37·200	5. 38·95
Y	3. 46·900	5. 52·50
Z	3. 41·700	6. 2·75
K	4. 19·100	5. 21·40
I	4. 34·300	5. 43·40
⊙	10. 5. 2·733	10. 6. 15·25
	3. 34·383	4. 40·65

...tact 5m 19.0 ✓

...out at 11h 18m Green...
two patches A & B
...inutes, till 11.20
... the usual dark,
& position, being
11.23. The details
...ond the RA handle.)

1859. September 1. Thursday.

Fair, with passing clouds. An extraordinary occurrence was witnessed which is detailed below. Too busy cutting trees to watch for a repetition.

10.33.0 App = 10.38.55.0 Chr
Bar 29.37 Th 60.5

Bar A

☉	9.51.6.1	10.0.6.0	10.15.9.3
V	52.12.3	1.11.9	16.12.0
W	–	1.31.7	–
X	–	1.37.2	–
Z	–	1.41.7	–
Y	–	1.46.9	–
P	52.27.1	1.27.0	16.27.1
r	52.29.3	–	16.28.7
R	52.37.2	–	16.36.8
S	52.42.3	–	16.42.1
Q	52.47.4	–	16.46.9
U	52.47.8	–	16.47.7
T	52.54.1	1.53.9	16.53.7
K	53.19.1	2.19.2	17.19.0
I	53.34.4*	2.34.2	17.34.3
☉	9.54.2.4	10.3.2.9	10.18.2.9

Bar B

☉	9.55.6.1	10.11.6.0
Q	55.48.7	11.48.3
P	55.53.4	11.53.1
R	56.1.1	12.0.5
r	56.8.1	12.8.2
S	56.8.4	12.8.4
T	56.12.1	12.11.3
U	56.20.8	12.21.1
V	56.58.8	12.58.4
K	57.21.4*	13.21.4
W	57.27.2	13.26.9
X	57.39.1	13.38.8
Z	57.43.5	13.43.3
Y	57.54.6	13.54.4
Z	58.2.9	14.2.6
☉	9.58.15.2	10.14.15.3

SAGGIO DEI DISEGNI GIORNALIERI DELLA CROMOSFERA SOLARE PRESICOLLO SPETTROSCOPIO

Photograph by Lorenzo Respighi, Royal Observatory of the Campidoglio, Rome, 1872
440 × 500 mm
Science Museum Group. Object no. 1876-1470/5

This rather strange image, reminiscent of a musical score, is a compilation of photographs of 28 drawings of the Sun, produced by the Italian astronomer Lorenzo Respighi during the spring of 1872.

Each line shows the edge of the Sun's disc on a single day, unrolled into a straight line, as Respighi tracked the appearance and disappearance of prominences in the solar atmosphere. Respighi made pioneering use of spectroscopic techniques to produce his images, allowing him to view the Sun in a single colour of light and see details that would normally be hidden behind its dazzling glare.

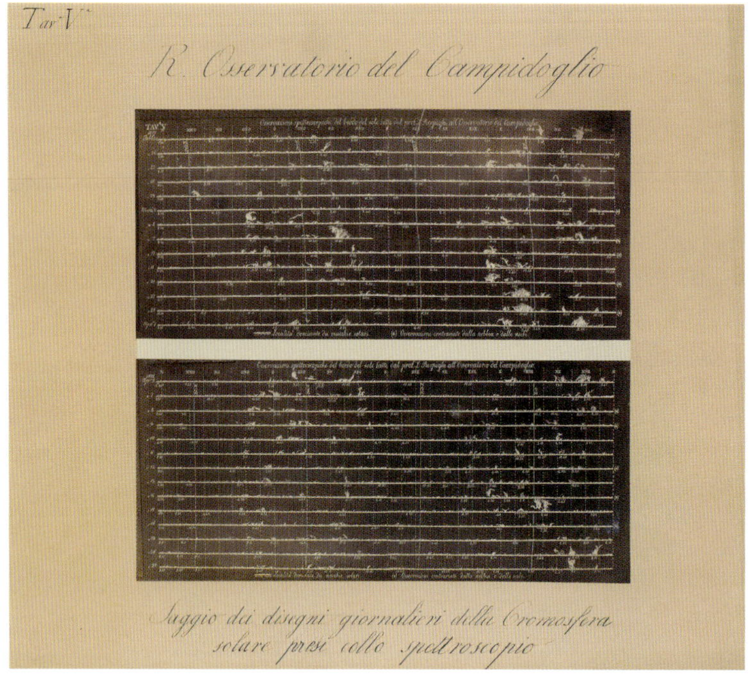

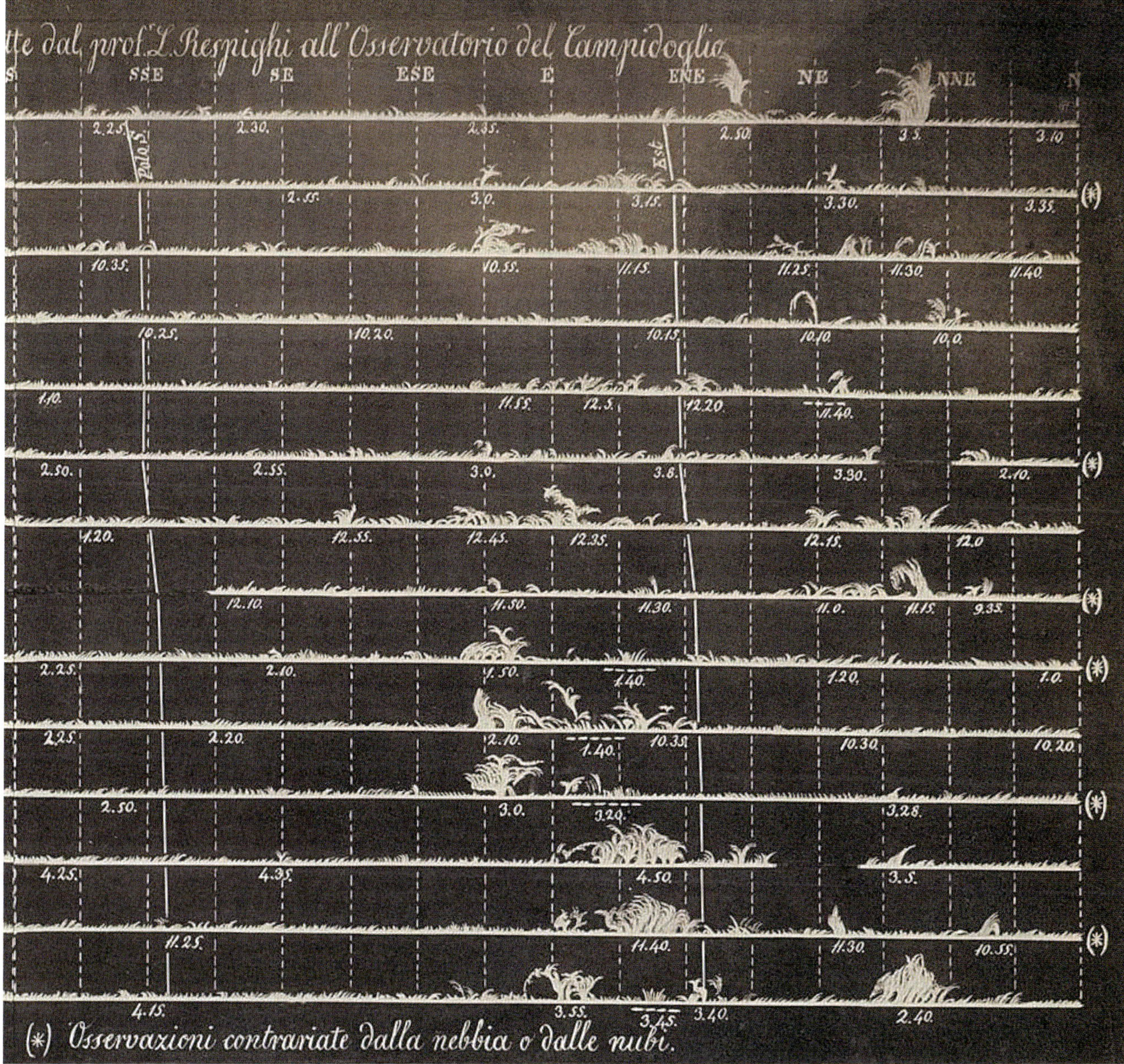
tte dal prof. L. Respighi all'Osservatorio del Campidoglio
S
SSE
SE
ESE
E
ENE
NE
NNE
N
Polo S.
Est
(*) Osservazioni contrariate dalla nebbia o dalle nubi.

ERUPTIVE PROMINENCES

Woodcut print from Samuel Pierpont Langley's
The New Astronomy, Boston, 1888
250 × 176 mm (manuscript page)
British Library

As secretary of the Smithsonian Institute in Washington, DC, Samuel Pierpont Langley founded both the Astrophysical Observatory and the National Gallery of Art. He had excellent drafting skills and, dissatisfied with the quality of early solar photography, turned to drawing to record his observations of the Sun's surface. His book *The New Astronomy* started as a series of lectures and was serialised in the magazine *Century*. He sought to gather public support for 'the new astronomy', or what we now call solar physics.

The first two chapters are devoted to the Sun's surface and emissions. Langley used 47 illustrations, both his own and those of others, to demonstrate the changing understanding of these phenomena, which he specifically compared to images of known earthly forms, like clouds and crystals. The print here, copied from his fellow-American Charles Young, shows eruptive prominences, likened variously to hedgehog spines, sheaves of grain and whirling waterspouts.

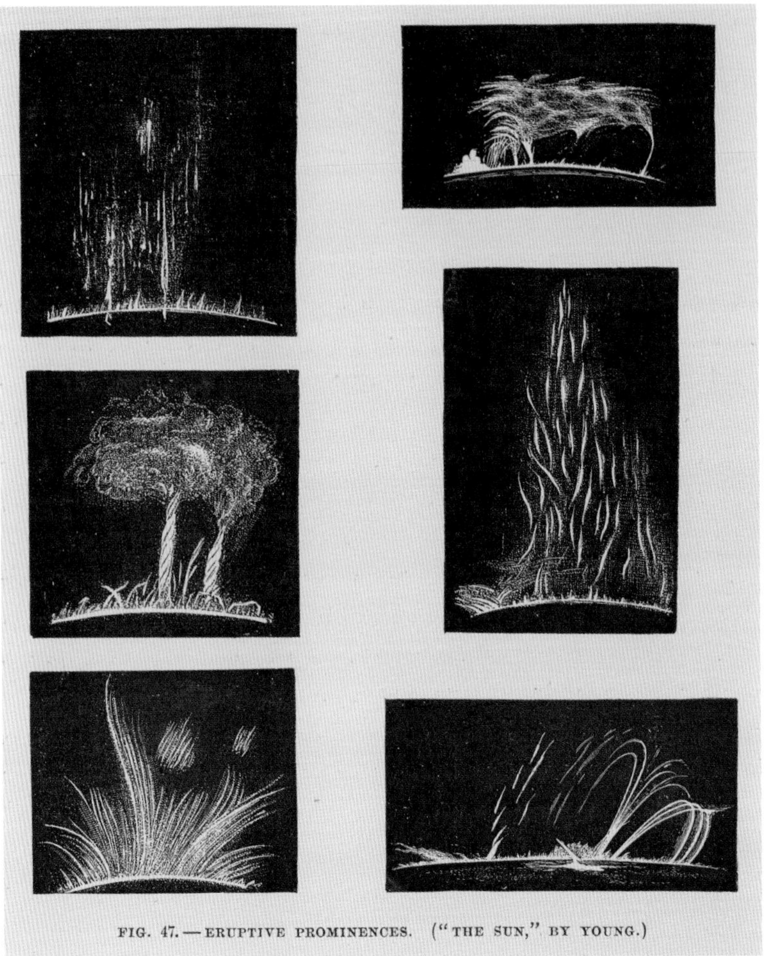

FIG. 47.—ERUPTIVE PROMINENCES. ("THE SUN," BY YOUNG.)

ERUPTING PROMINENCE

Glass positive spectroheliogram by John Evershed,
Srinagar, Kashmir, 26 May 1916
250 × 200 mm
Science Museum Group Object no. 1929-716

This photograph by the astronomer John Evershed captures an enormous eruption of material from the Sun. Evershed spotted the start of the eruption at 7.47 am from his observing station in Kashmir, taking a series of photographs over a period of around two hours, as he tracked the vast cloud of plasma rising into space at 300 km per second.

At its greatest extent, the prominence towered 800,000 km above the solar surface, making it the highest on record at the time. Evershed described it as appearing like a 'fountain', which later faded and broke apart like 'a series of beads strung on an invisible cord'.*

* John Evershed, 'The highest prominence on record', *Journal of the Astronomical Society of India*, 7, nos 7–9 (1917), pp. 41–4.

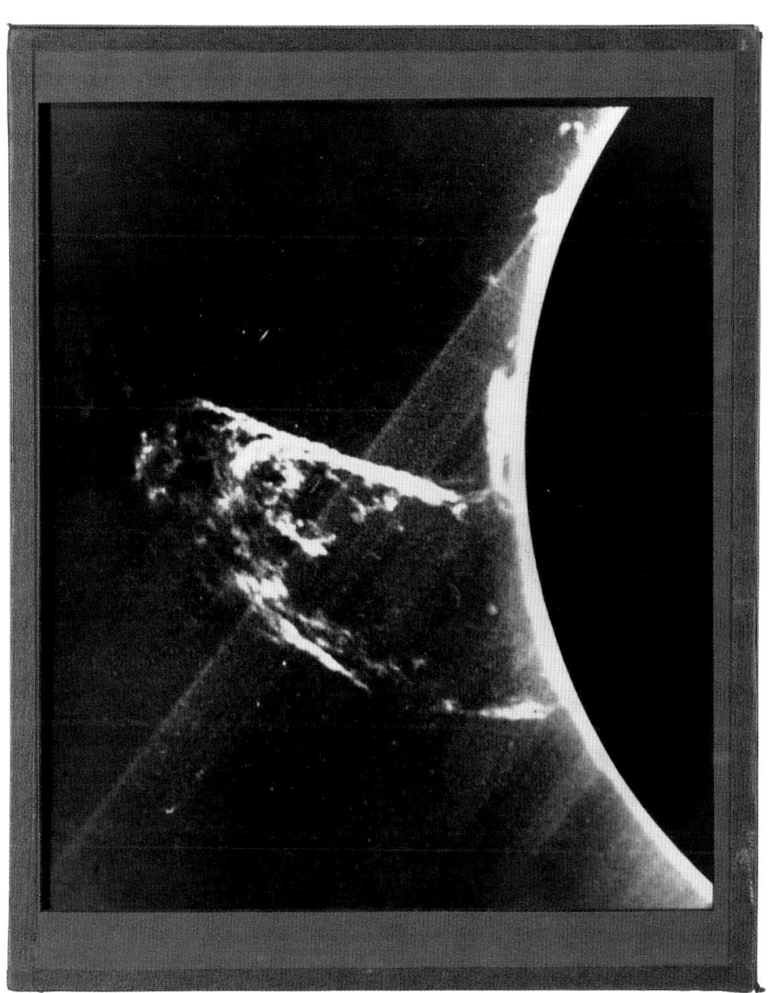

ELBOW PROMINENCE

Photograph in extreme ultraviolet by Owen Garriott, Skylab space station, 9 August 1973
Digital image
NASA

Some prominences hang lazily above the surface of the Sun like clouds, before suddenly erupting away from the surface at tremendous speed, driven by some unseen force. Such eruptions are thought to be triggered by the violent release of stored magnetic energy, as twisted magnetic loops close to the surface of the Sun snap and reconnect.

The astronaut Owen Garriott took this photograph from Skylab, the USA's first crewed space station. Multiple fainter images of the Sun are visible across the frame, showing it as it appears in different wavelengths. Beyond the obscuring effects of the Earth's atmosphere, space-based solar astronomy is able to provide stunning new vistas on our nearest star.

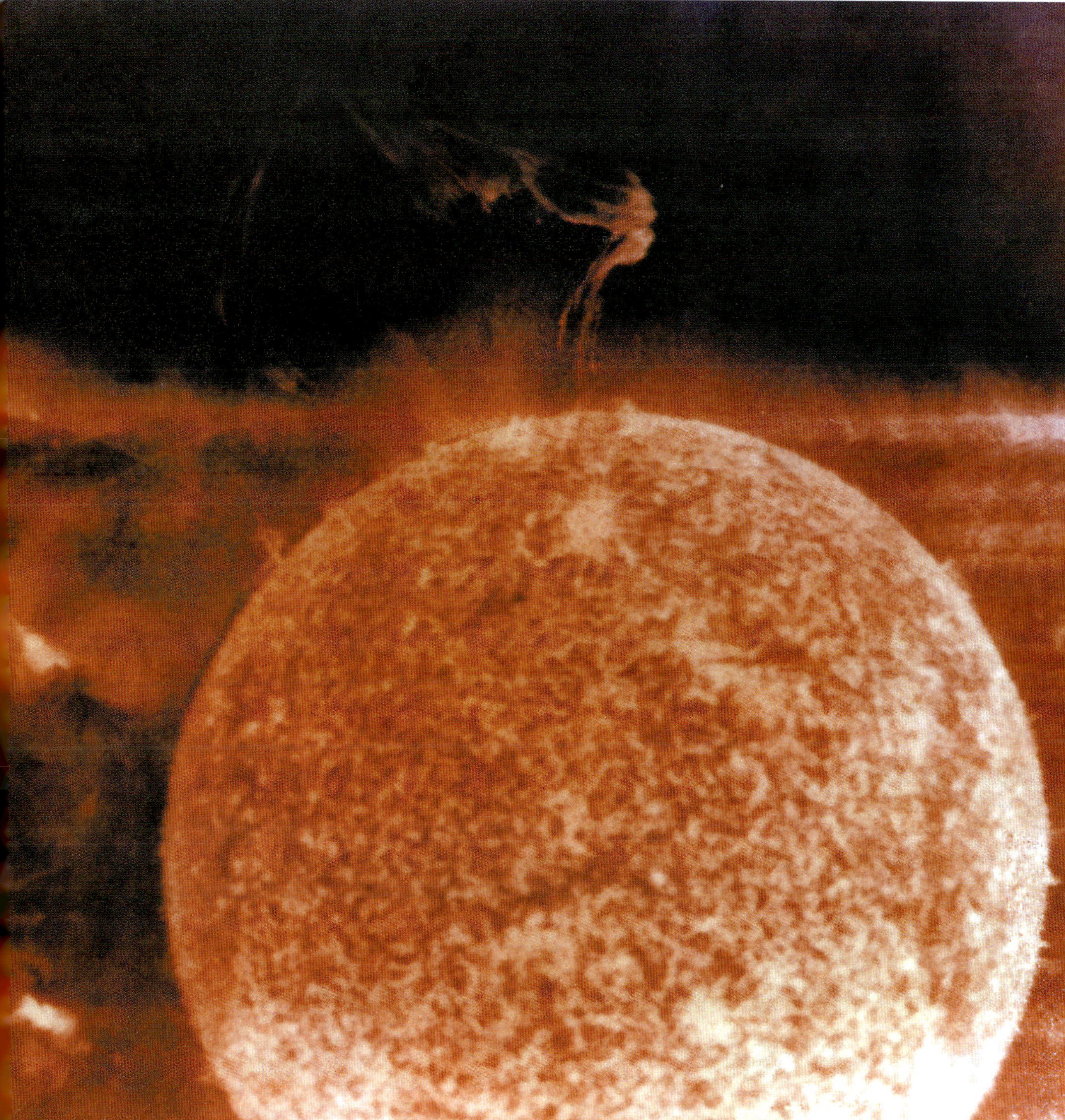

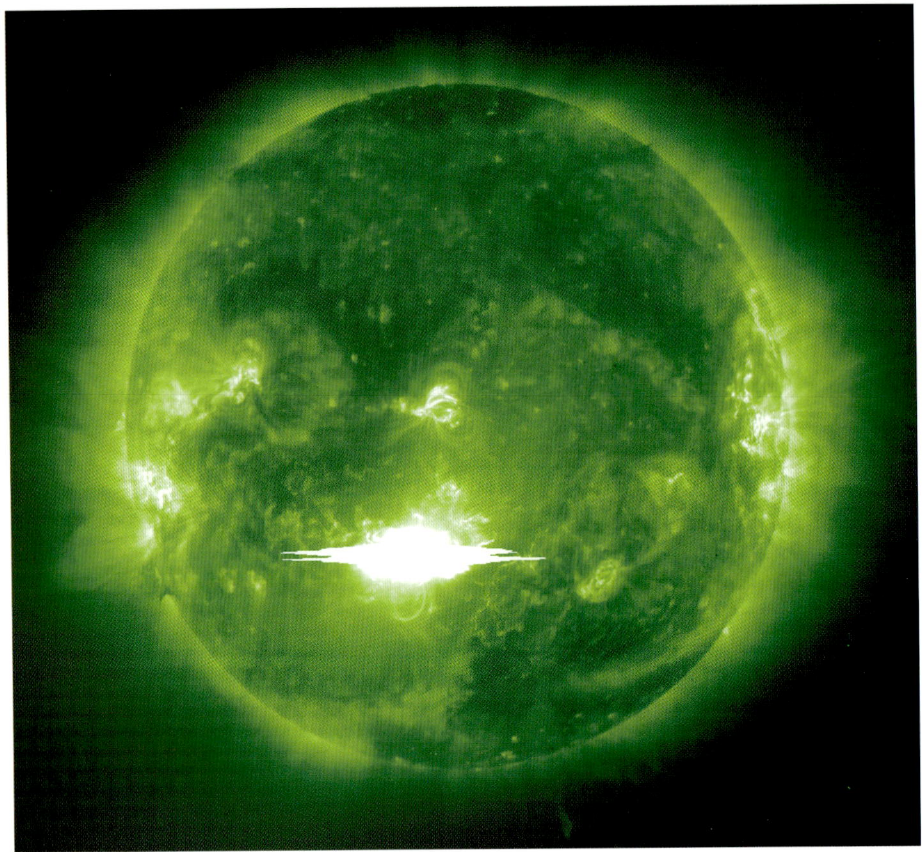

SOLAR FLARE

Photograph in extreme ultraviolet by the Solar and Heliospheric Observatory (SOHO), 28 October 2003
Digital image
European Space Agency (ESA) and NASA

This image captures a solar flare, a brilliant flash of electromagnetic radiation that blasted the Earth with X-rays. The lurid green colour was added artificially to help astronomers identify the wavelength of ultraviolet light used to produce the image. This flare was part of a series of unusually violent solar events known as the Halloween Storms. Huge clouds of electrically charged gas, or coronal mass ejections, struck the Earth, disrupting communications, diverting planes, knocking out electrical power in Sweden and creating auroras that were visible in Texas and the Mediterranean. Even the spacecraft that took this photograph failed temporarily.

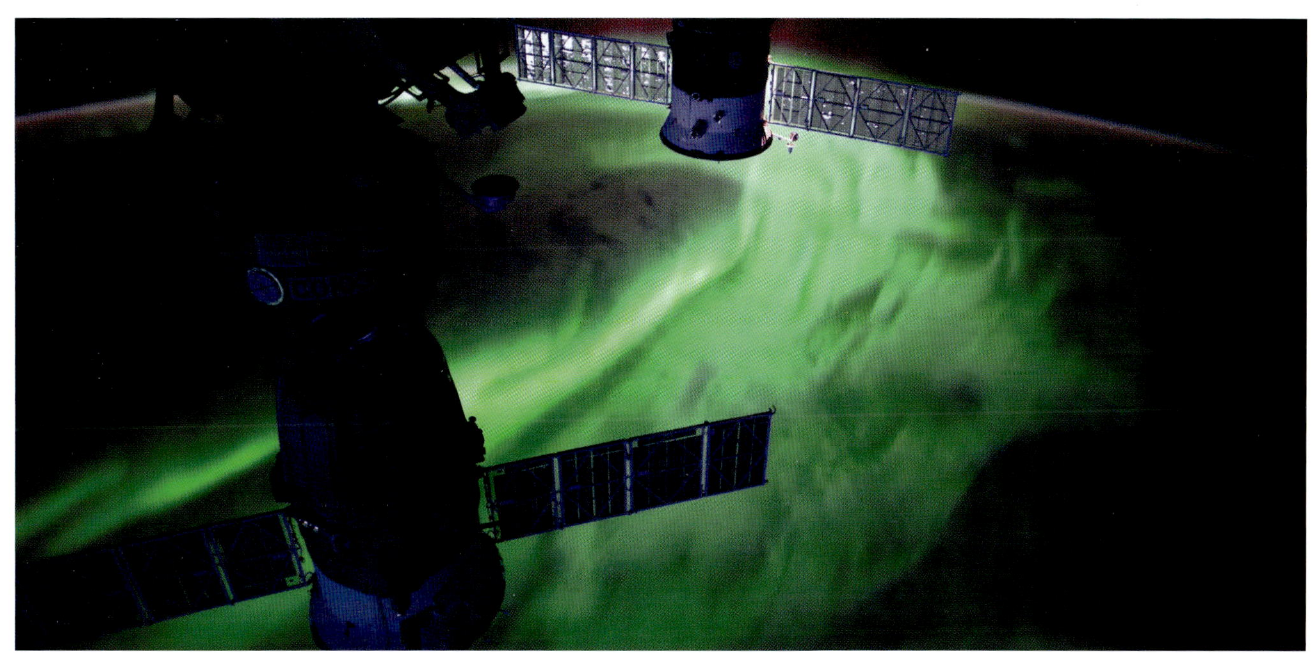

AURORA FROM SPACE

Photograph taken from the International Space Station, 19 April 2016
Digital image
NASA

Taken from the International Space Station (ISS), this photograph of an aurora is a spectacular reminder of the fact that we live in the Sun's atmosphere. Two Russian Soyuz spacecraft, docked with the ISS, frame the green swirl of the aurora.

Our local star sends a constant stream of electrically charged particles hurtling through space at up to 750 km per second. The Earth's magnetic field shields us from the worst of this solar wind, channelling it towards the poles where solar particles collide with our atmosphere, creating the otherworldly lights we know as the aurora. Auroras are most common in the Arctic or Antarctic but during powerful solar storms they can appear at much lower latitudes. Astronauts on the ISS take shelter in specially shielded compartments during such storms, to avoid potentially harmful doses of radiation.

CORONAL LOOPS

Photograph in extreme ultraviolet by NASA's Solar Dynamics Observatory (SDO), 15 January 2012
Digital image
NASA

The dancing curves arcing out of the Sun's surface are known as coronal loops. When material from the corona cools it can fall back towards the solar surface like rain and is channelled by strong magnetic fields into glowing threads of hot plasma.

These beautiful structures reveal the magnetic fields that curve across the Sun's surface, often connecting sunspots with opposite magnetic poles. Solar physicists believe that coronal loops are key to solving the mystery of why the corona is so much hotter than the solar surface. They are often found in regions that produce explosive outbursts of magnetic energy, in the form of flares and coronal mass ejections.

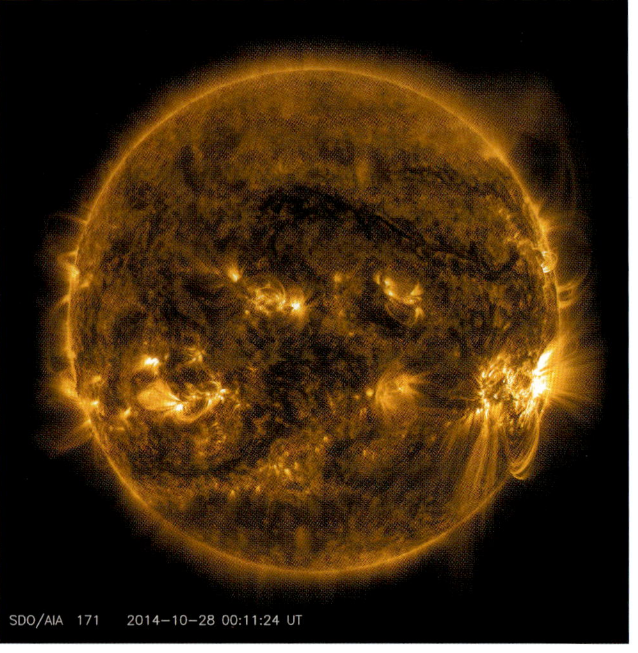

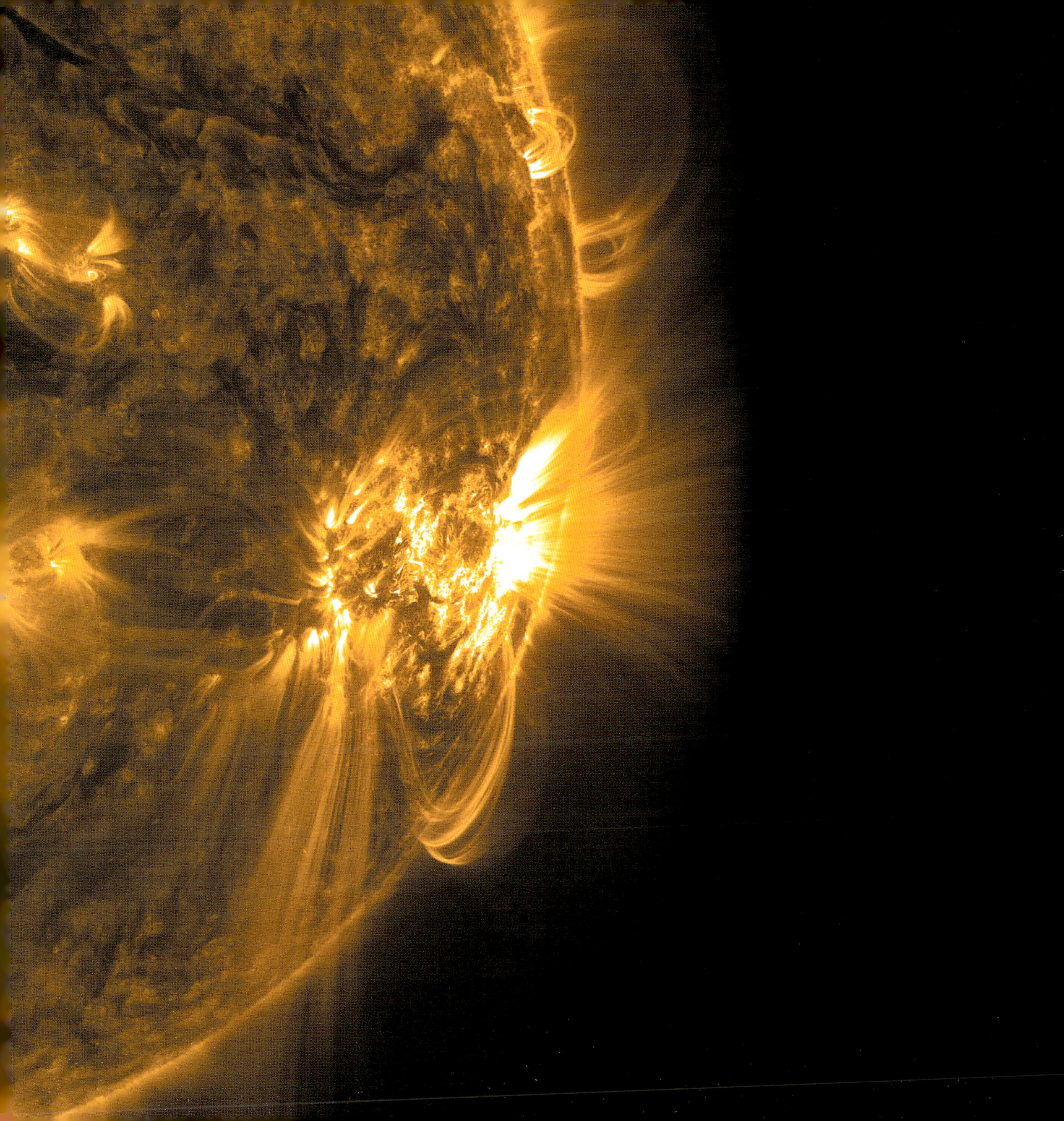

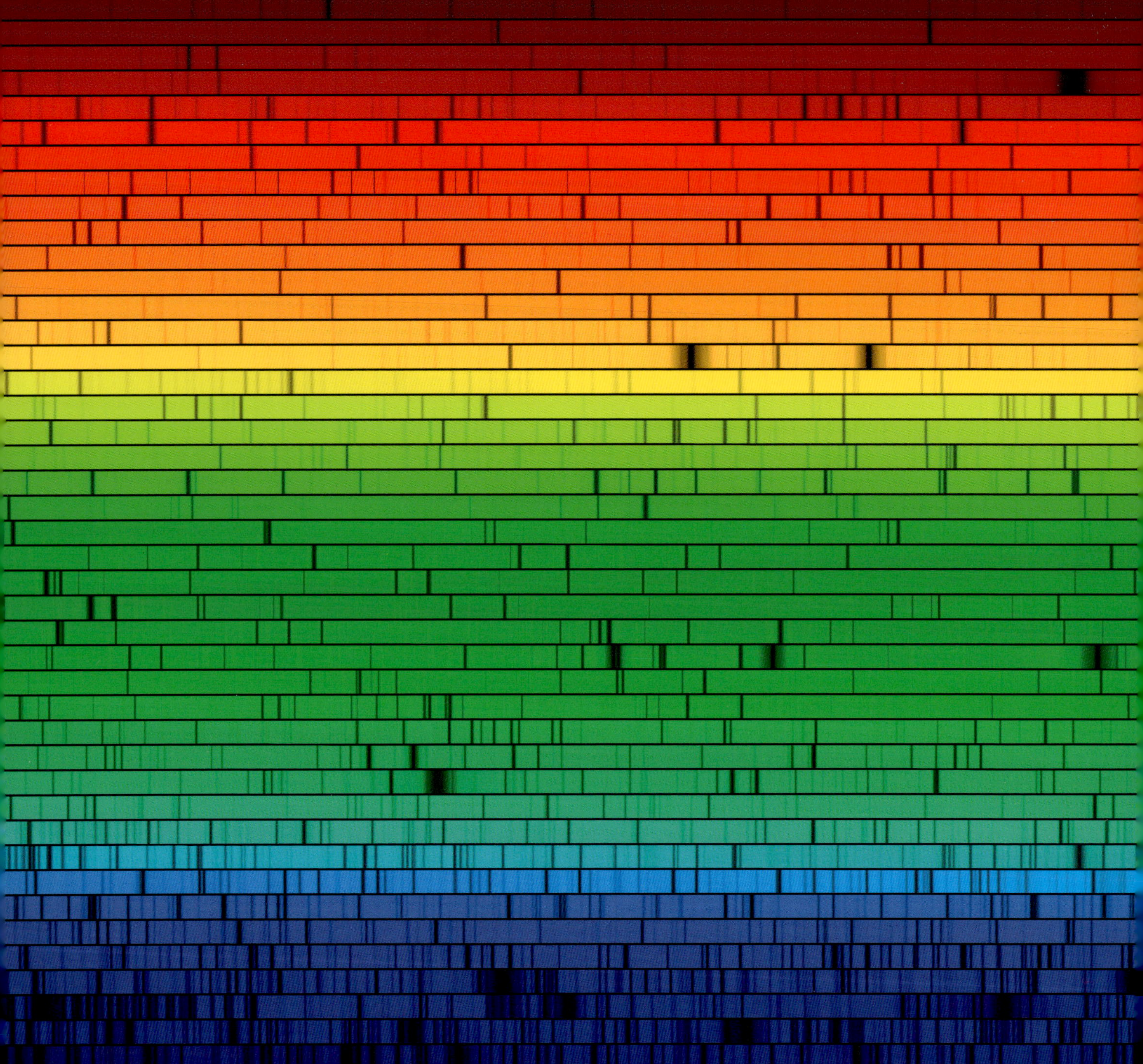

4

SUNLIGHT

To the human eye, sunlight appears white and uniform, yet there is a wealth of information encoded in its rays. In the 17th century, Isaac Newton performed experiments showing that sunlight was made up of a spectrum of seven colours, rather than acquiring colour when it passed through prisms or reflected off objects. At the start of the 19th century, William Hyde Wollaston examined the solar spectrum in detail and discovered a number of mysterious dark lines crossing the bands of Newton's colours. Although he did not understand them at the time, these dark lines would be key to unlocking the secrets of the Sun.

Wollaston's discovery heralded a completely new branch of science: spectroscopy. During the 19th century, work by physicists and chemists revealed that spectral lines were the unique fingerprints left by different chemical elements. The Sun had seemed so far beyond the reach of experiments that its true nature would never be fully understood. Now it could be investigated through this brand-new type of astronomy. Norman Lockyer declared that using spectroscopy he would 'take the very Sun itself to pieces'.*

By analysing spectra, astronomers discovered that the Sun contained elements familiar to chemists on Earth, and uncovered a wholly new one: helium. The invention of the spectroheliograph allowed the Sun to be viewed, and later photographed, in a single colour of light, revealing our nearest star as never seen before. Today space-based instruments produce images using light outside of the visible spectrum, in ultraviolet and X-rays, showing previously invisible details of our dynamic Sun.

* See Iwan Rhys Morus, *When Physics Became King* (Chicago: University of Chicago Press, 2005), p. 215.

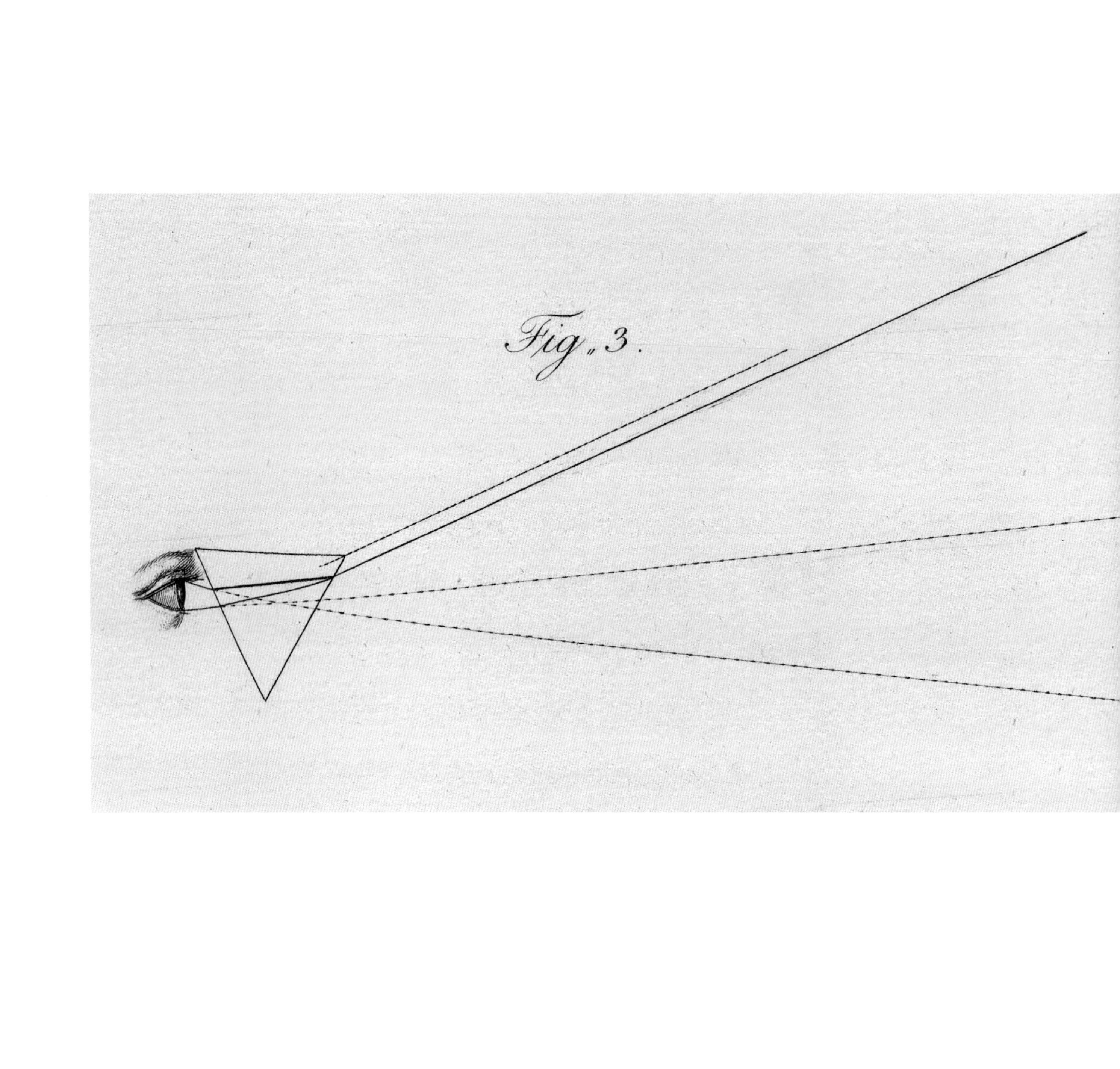
Fig. 3.

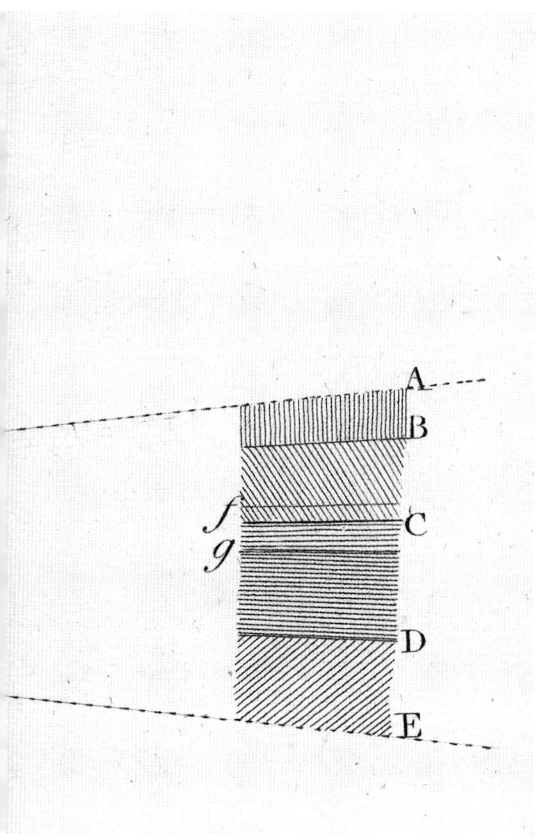

A METHOD OF EXAMINING REFRACTIVE AND DISPERSIVE POWERS, BY PRISMATIC REFLECTION

Engraved print from a paper by William Hyde Wollaston published in *Philosophical Transactions of the Royal Society*, LXXXXII, London, 1802
195 × 380 mm (manuscript page)
British Library

William Hyde Wollaston was investigating the ways in which different substances refract and disperse light when he made an unexpected discovery: a series of dark lines crossing the rainbow spectrum of sunlight. He concluded that these were the boundaries between colours, and that there were only four compared to Newton's claim of seven: red, yellowish green, blue and violet. Unbeknown to Wollaston, he had stumbled upon the key that would eventually allow scientists to discover the make-up of the Sun.

Faced with the question of how to visualise his findings in a black and white line engraving, he produced the illustration here. The lines are designated with the letters A to E, and two further lines that he saw within the colours as f and g. The colours are differentiated by patterns of lines in different directions. The limitations of this method are clear when you consider how the line for g is lost in the pattern for the blue spectrum between C and D.

Previous page
Detail of a spectrogram showing the visible solar spectrum, 1984 (see p. 93).

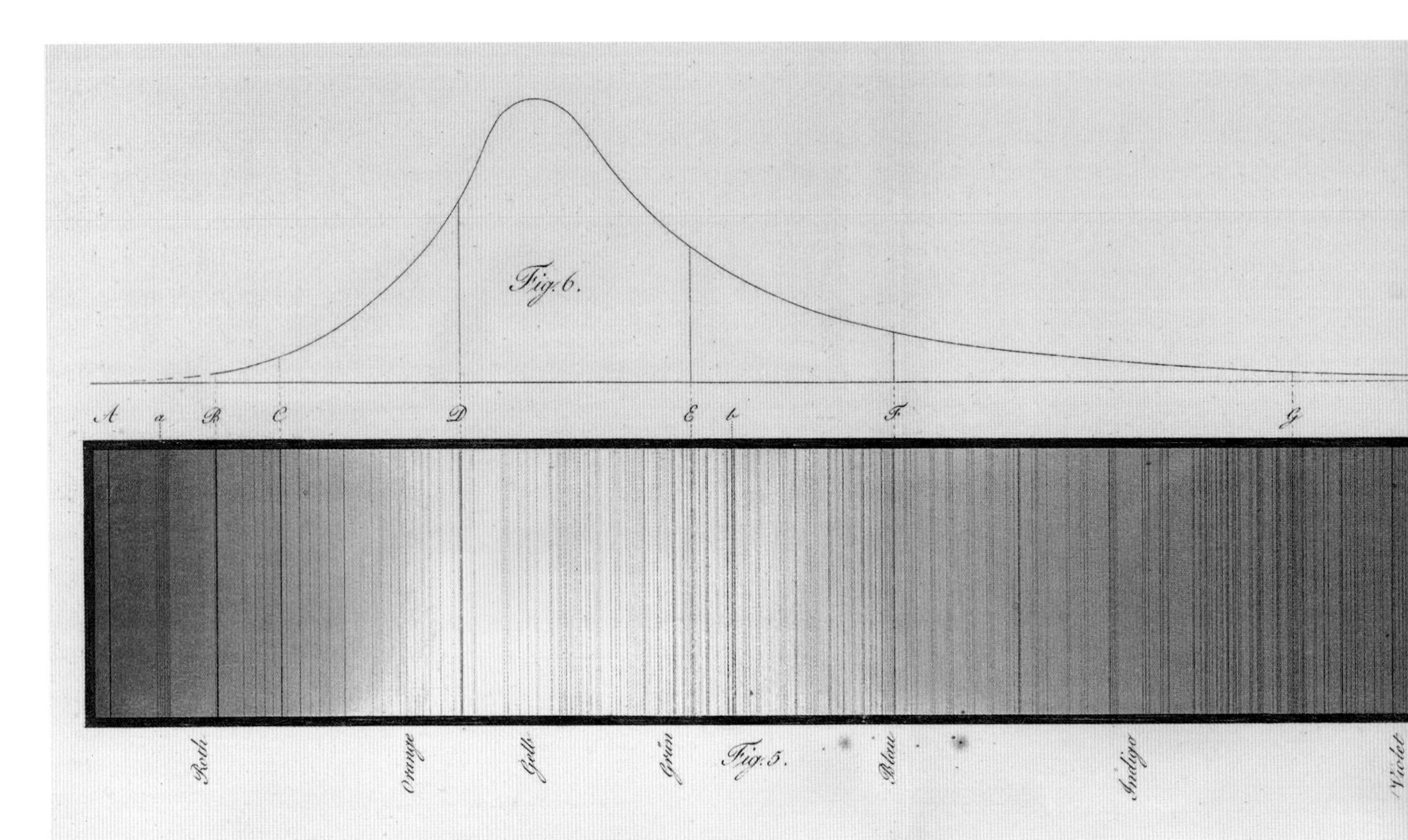

Zu Fraunhofer's Abh._Denkschr. 1814_15.

DETERMINATION OF THE REFRACTION AND COLOUR DISSEMINATION PROPERTY OF DIFFERENT TYPES OF GLASS

Etching with wash in Indian ink by Joseph von Fraunhofer for his paper published in *Denkschriften der Königlichen Akademie der Wissenschaften München*, vol. 5, Munich, 1814–15
258 × 430 mm
Science Museum Group. Object no. Q TM 957

Joseph von Fraunhofer was the first person to map comprehensively the dark lines found in the solar spectrum, which were later named after him. Bringing together a combination of craftsmanship and theory, he sought originally to improve the production of optical lenses for instruments. In communicating his findings to the Bavarian Academy of Sciences, Fraunhofer went to great lengths to provide an exact image of what he had seen. He etched the printing plate for this figure himself, carefully matching the corrosion of each metal line to the intensity of the spectral line, and adding a wash of ink to increase the intensity at either end of the spectrum. He mapped 374 different lines between B at the red end of the spectra and H at the violet end. Some rare hand-coloured copies of this figure exist, but Fraunhofer avoided this expense by applying exquisite attention to his etched lines.

RADIATION AND ABSORPTION SPECTRA

Lithograph from *Studies in Spectrum Analysis* by Norman Lockyer, London, 1878
183 × 328 mm
Science Museum Group. Object no. 540 LOCKYER

Norman Lockyer, the first director of the South Kensington Solar Physics Observatory, was particularly excited by the possibilities of the new field of spectroscopy. His *Studies in Spectrum Analysis* was intended to demonstrate how studying spectra could allow scientists to determine the make-up of astronomical bodies. This lithographic plate compares spectra produced by known substances in a laboratory to the spectrum of sunlight.

Spectrum 8 displays the presence of sodium, magnesium and hydrogen in the Sun's chromosphere by comparison to their laboratory spectra (2, 3 and 6 respectively). Spectrum 9 shows absorption in the solar spectrum and how the strong double lines visible in the yellow region correspond to those of sodium in spectrum 10. In 1868, Lockyer and the French astronomer Jules Janssen had found a new spectral line close to these double lines of sodium. Lockyer realised that this was evidence of a new element, which he named helium after Helios, the ancient Greek personification of the Sun. Helium was the first element to be discovered in space and was not isolated on Earth until 1895.

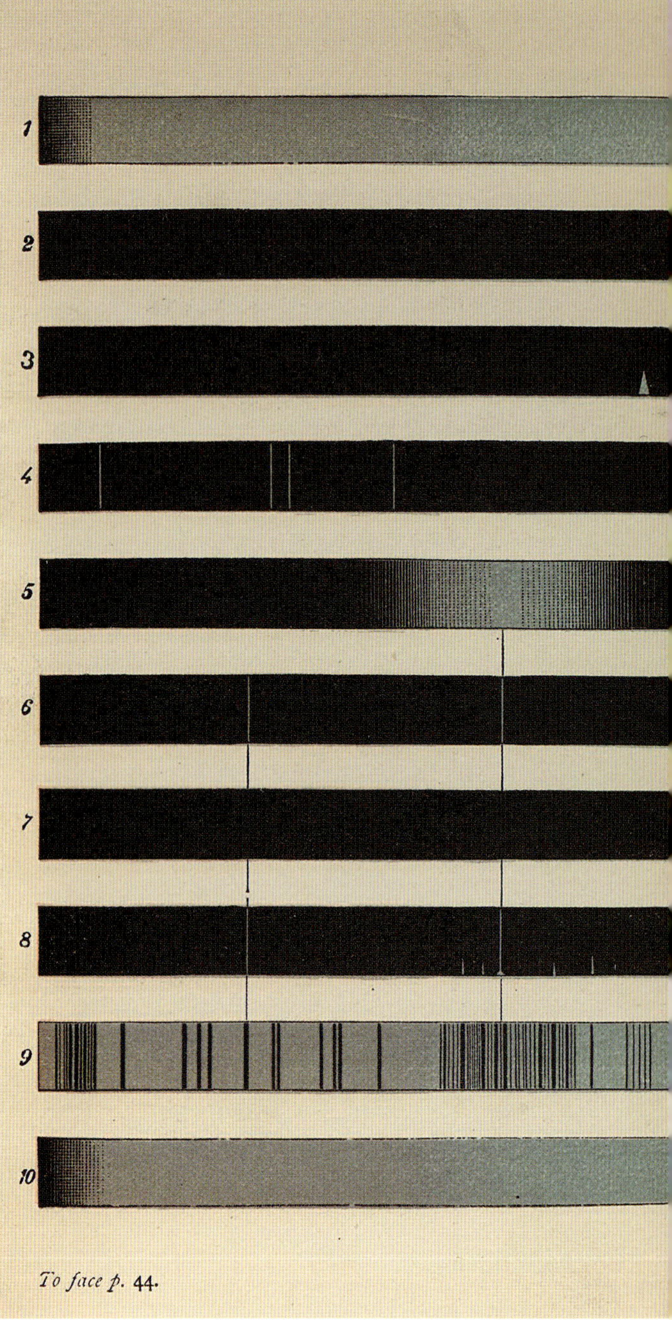

Plate II.

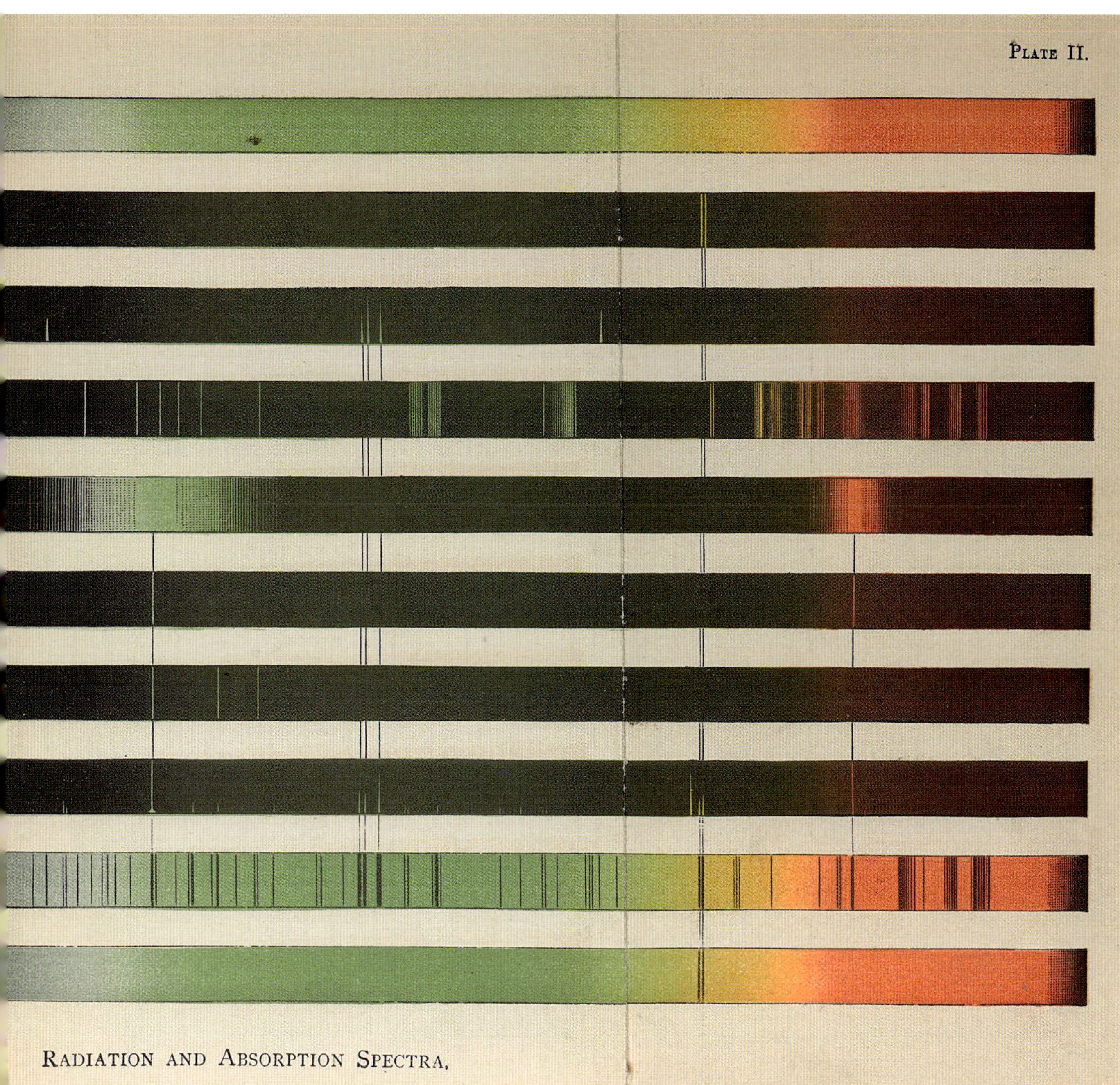

Radiation and Absorption Spectra.

DRAPER'S DAGUERREOTYPE OF THE SOLAR SPECTRUM

Daguerreotype on yellow iodide of silver by
John William Draper, Virginia, USA, 27 July 1842
405 × 305 mm
Science Museum Group. Object no. 1948-316

John William Draper was a professor of chemistry and natural philosophy at the University of the City of New York, and a pioneer of photography. He captured some of the first daguerreotype photographic images of the spectrum of sunlight and sent this one to the famous English astronomer, John Herschel.

Draper compared his photographs to Fraunhofer's map of the dark lines in the solar spectrum, but in this image, at least, he failed to find any in the yellow, orange and green parts of the spectrum (see p. 86). However, he did discover new spectral lines in the infrared and ultraviolet, which are invisible to the human eye but revealed by the application of photography. He has drawn directly into the photographic plate to mark his spectral lines.

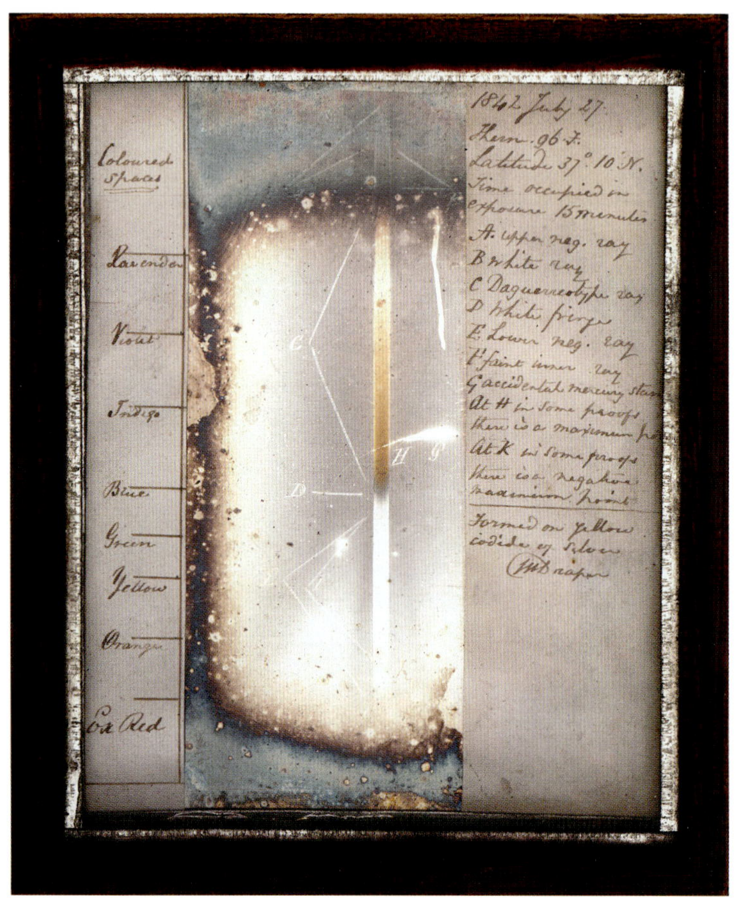

SPECTRAL LINES OF CALCIUM IN SUNSPOTS

Glass positive spectrogram by John Evershed,
Kodaikanal Observatory, India, *c*.1909
253 × 253 mm
Science Museum Group. Object no. EVER/A/1/E119

The English astronomer John Evershed first visited the Kodaikanal Observatory in India in 1907 and spent 16 years there observing the Sun. This seemingly abstract image, taken using a spectrograph, shows the dark spectral lines around a group of sunspots (top) and a single sunspot (bottom). The vertical lines are caused by calcium vapour in the Sun's atmosphere, whereas the dark horizontal bands indicate the positions of the sunspots.

Evershed noticed that the spectral lines seem to bend slightly where they cross the centre of a sunspot, suggesting that material is flowing out of the sunspot close to the surface of the Sun, and away from the sunspot higher in the solar atmosphere. This flow of material near sunspots became known as the 'Evershed effect'.

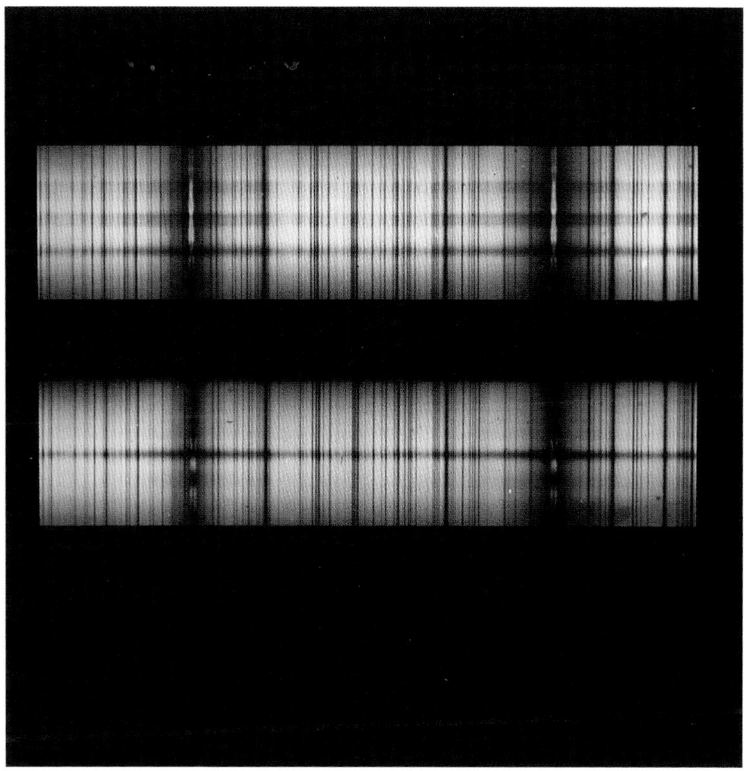

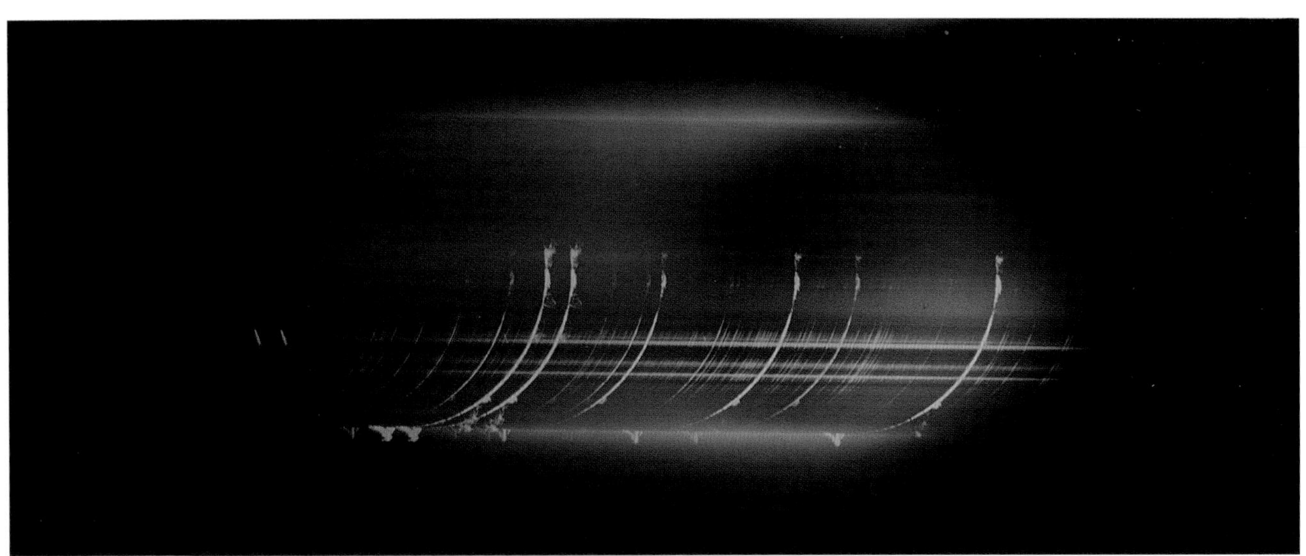

SPECTRUM OF THE CHROMOSPHERE DURING AN ECLIPSE

Glass positive spectrogram, Benkoelin, Sumatra, 14 January 1926
150 × 380 mm
Science Museum Group. Object no. 1999-911

Spectral lines in sunlight usually appear as dark bands against a bright background, as atoms in the solar atmosphere absorb certain colours of light. In this image, however, things are reversed.

The thin layer just above the visible surface of the Sun is known as the chromosphere due to its bright red colour, which is caused by superheated atoms emitting, rather than absorbing, certain colours. The chromosphere is usually too faint to see, but when the visible disc of the Sun is blocked out at the very start and end of a total eclipse this thin red layer can be seen just above the solar surface.

This image shows the spectrum produced by the chromosphere during the eclipse of 1926, creating multiple bright crescent-shaped lines with large prominences visible.

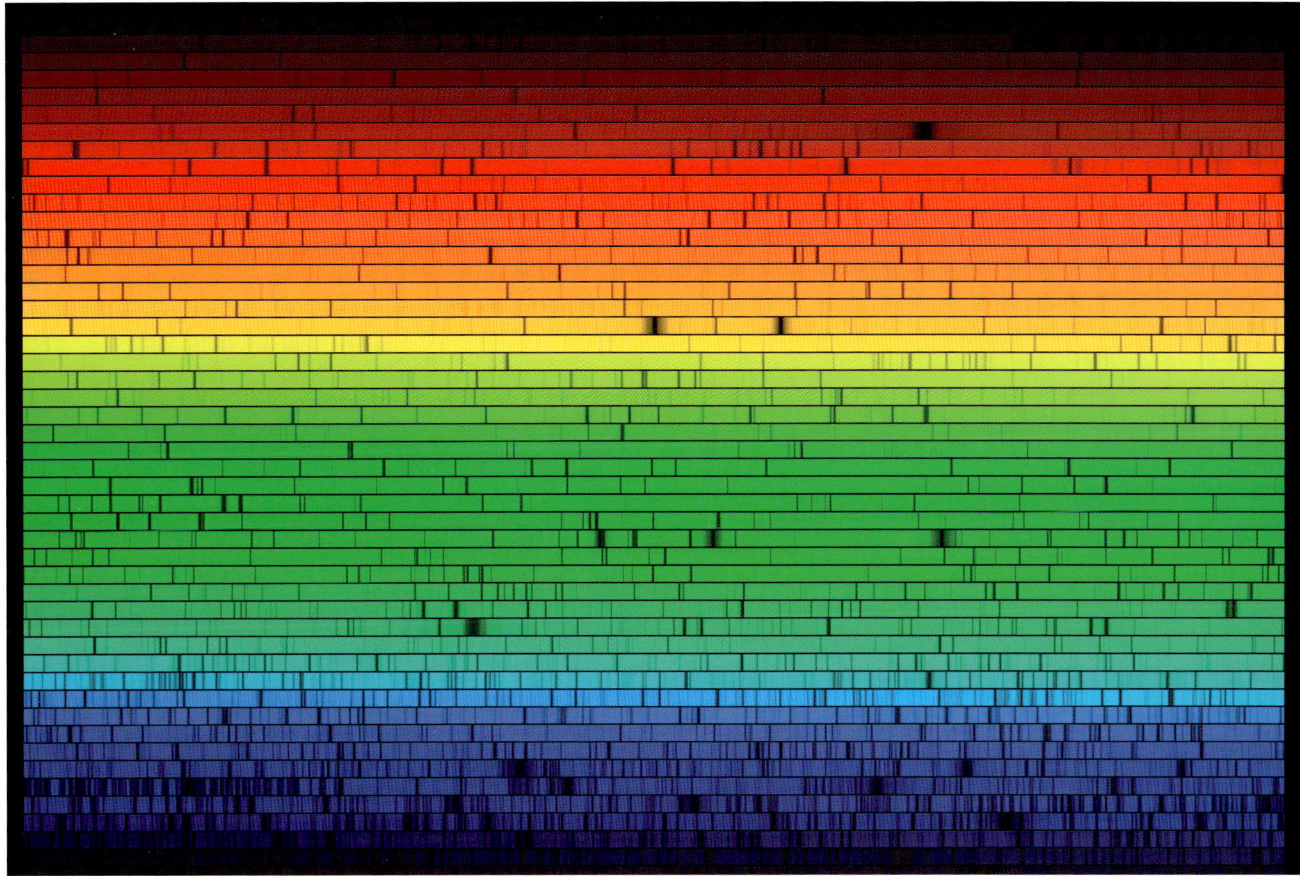

THE VISIBLE SOLAR SPECTRUM

Spectrogram made using the Fourier Transform Spectrometer, National Solar Observatory, Kitt Peak, Arizona, USA, 1 June 1984
Digital image
N A Sharp, NOAO/NSO/Kitt Peak FTS/AURA/NSF

This extraordinary rainbow shows the complete spectrum of visible sunlight, from red to violet. The bands of colour are crossed by thousands of dark spectral lines, produced as atoms in the solar atmosphere absorb characteristic colours of light.

This image contains a vast amount of information about the make-up of our Sun. Each chemical element has a distinguishing set of absorption lines, almost like a barcode that can be used to uniquely identify it. Spectroscopy has allowed scientists to determine that the Sun is made mostly of hydrogen and helium, and is a key tool in studying astronomical bodies that lie far beyond our reach, including distant stars and the planets in orbit around them.

THE SUN IN DIFFERENT WAVELENGTHS

Twelve photographs by NASA's Solar Dynamics Observatory (SDO),
28 December 2015
Digital image
NASA

These 12 images were taken within a few minutes of each other by NASA's Solar Dynamics Observatory, showing the Sun in various wavelengths of light, from visible to extreme ultraviolet. The wavelengths reveal different details of our dynamic star, from the tenuous corona to twisted coils of magnetic flux. In reality the images are monochrome, but astronomers allocate each wavelength a standard colour, so that they can quickly recognise the wavelength used to produce the image. Taking one ultra-high-resolution image of the Sun every second, the SDO orbits at a fixed point, 35,789 km above the surface of the Earth, allowing it to send large amounts of data in a continuous stream. The astonishing imagery that scientists produce from this data is essential to improving our understanding of our local star.

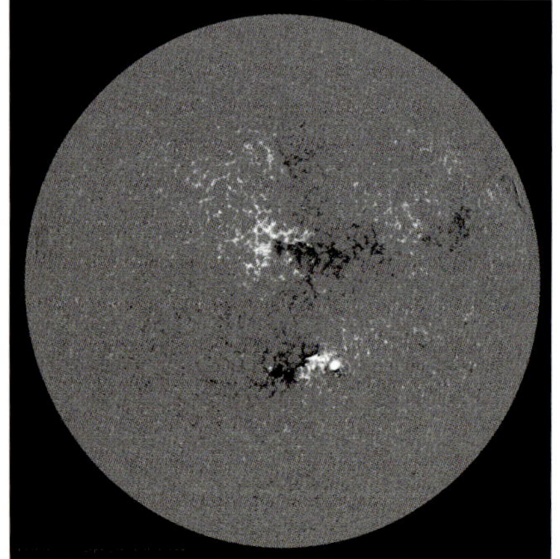

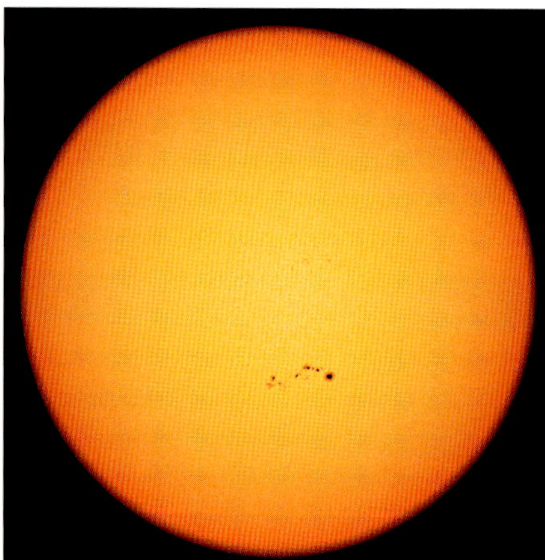

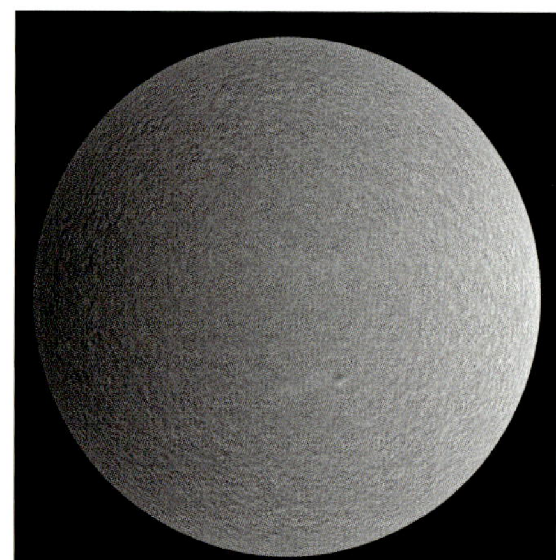

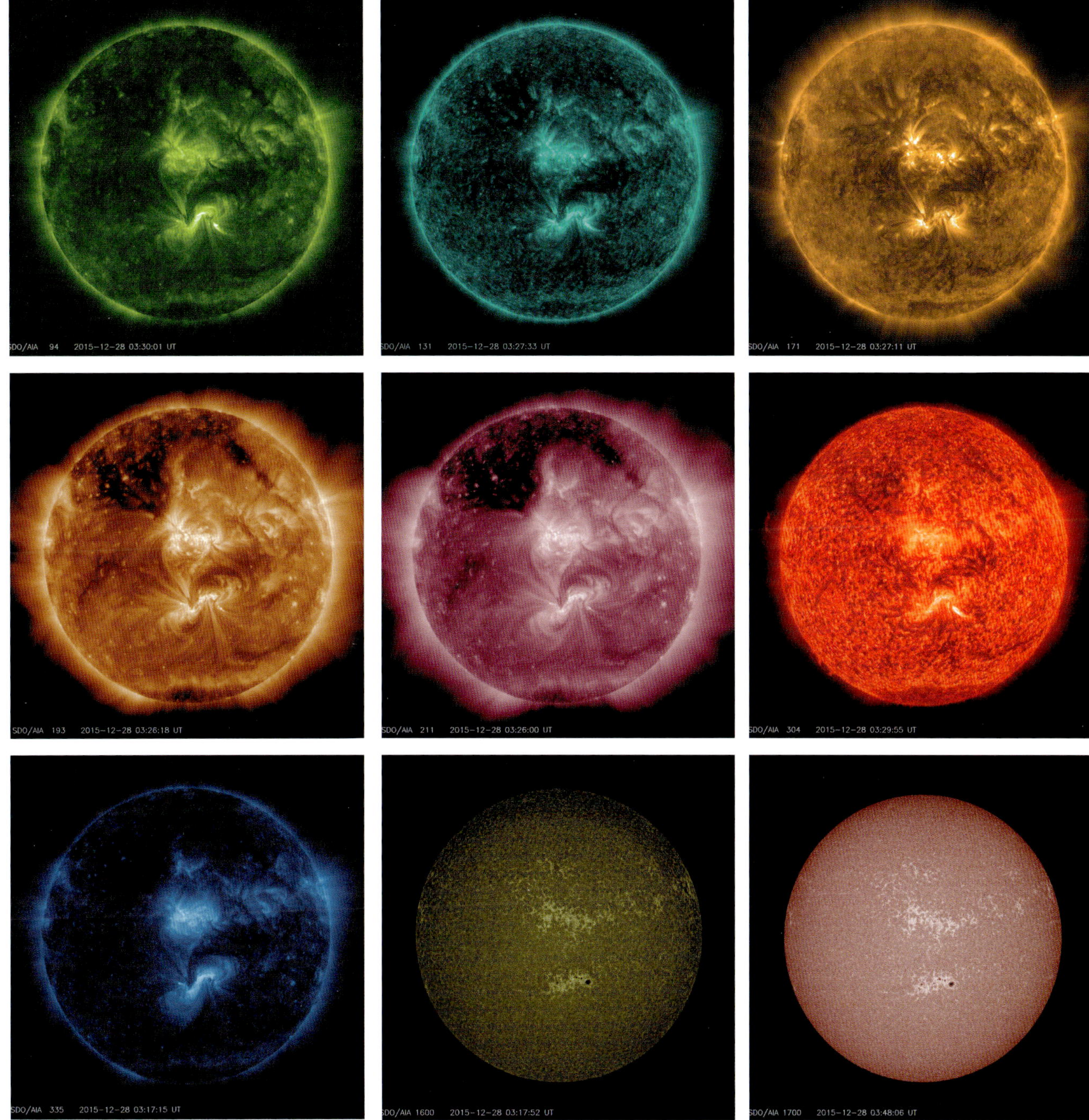
SDO/AIA 94 2015-12-28 03:30:01 UT
SDO/AIA 131 2015-12-28 03:27:33 UT
SDO/AIA 171 2015-12-28 03:27:11 UT
SDO/AIA 193 2015-12-28 03:26:18 UT
SDO/AIA 211 2015-12-28 03:26:00 UT
SDO/AIA 304 2015-12-28 03:29:55 UT
SDO/AIA 335 2015-12-28 03:17:15 UT
SDO/AIA 1600 2015-12-28 03:17:52 UT
SDO/AIA 1700 2015-12-28 03:48:06 UT

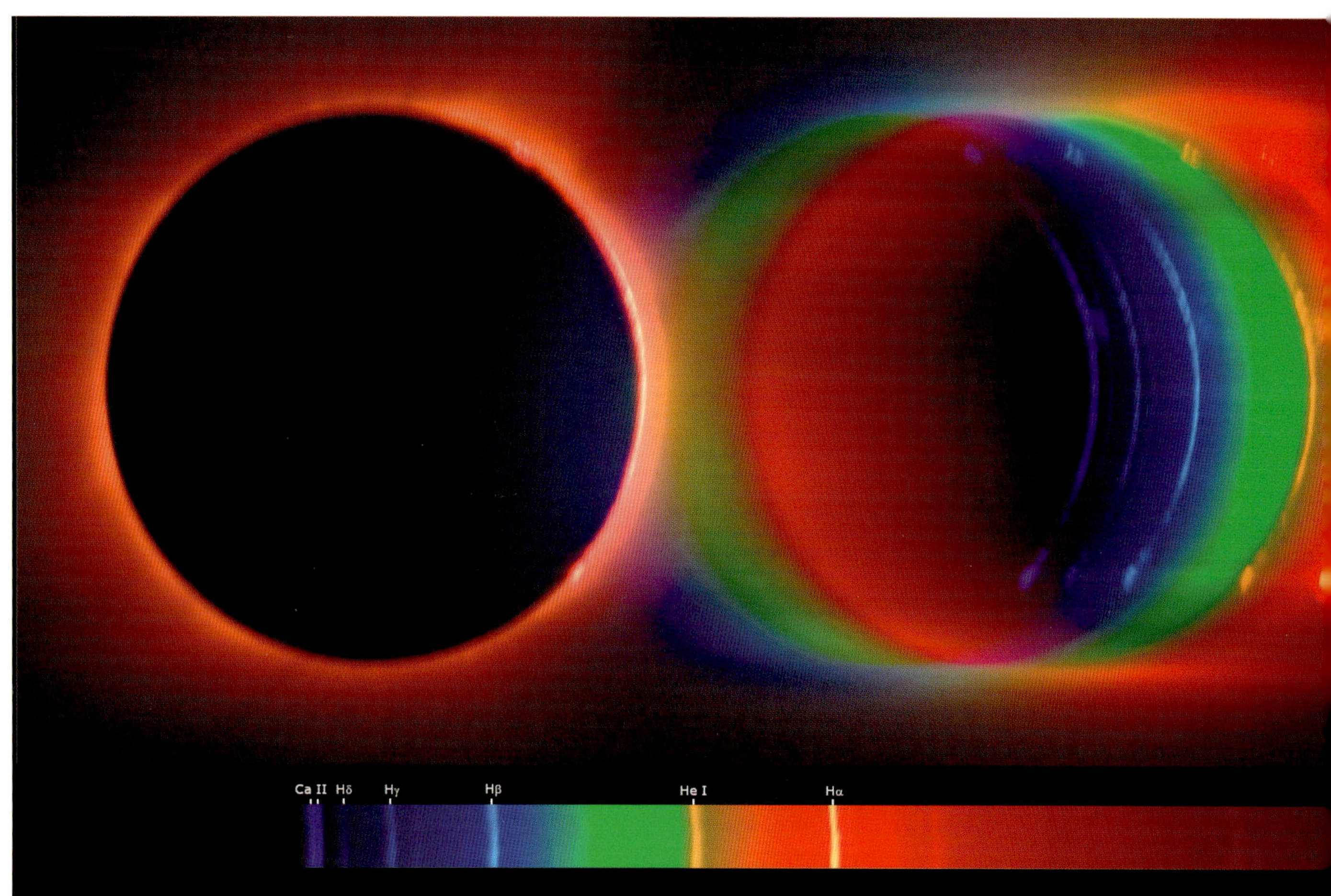
Ca II
Hδ
Hγ
Hβ
He I
Hα

THE COLOUR SPHERE OF THE SUN

Photograph taken by the CESAR (Cooperation through Education in Science and Astronomy Research) project, Casper, Wyoming, USA, 21 August 2017
Digital image
ESA

This image shows the bright colours of a 'chromosphere flash spectrum', taken during the solar eclipse of August 2017. The chromosphere is a relatively thin layer of the Sun's atmosphere that sits just above the much brighter surface, known as the photosphere. When the light of the photosphere is blocked by the Moon during an eclipse, the chromosphere becomes visible, although only for a few seconds at the very start and end of totality, so astronomers have to work fast to capture images: hence the name 'flash' spectrum.

The colours in this image show the unique fingerprints of different chemical elements. The strong red line is due to hydrogen, while yellow is the signature of helium, originally discovered in a flash spectrum studied during the eclipse of 1868.

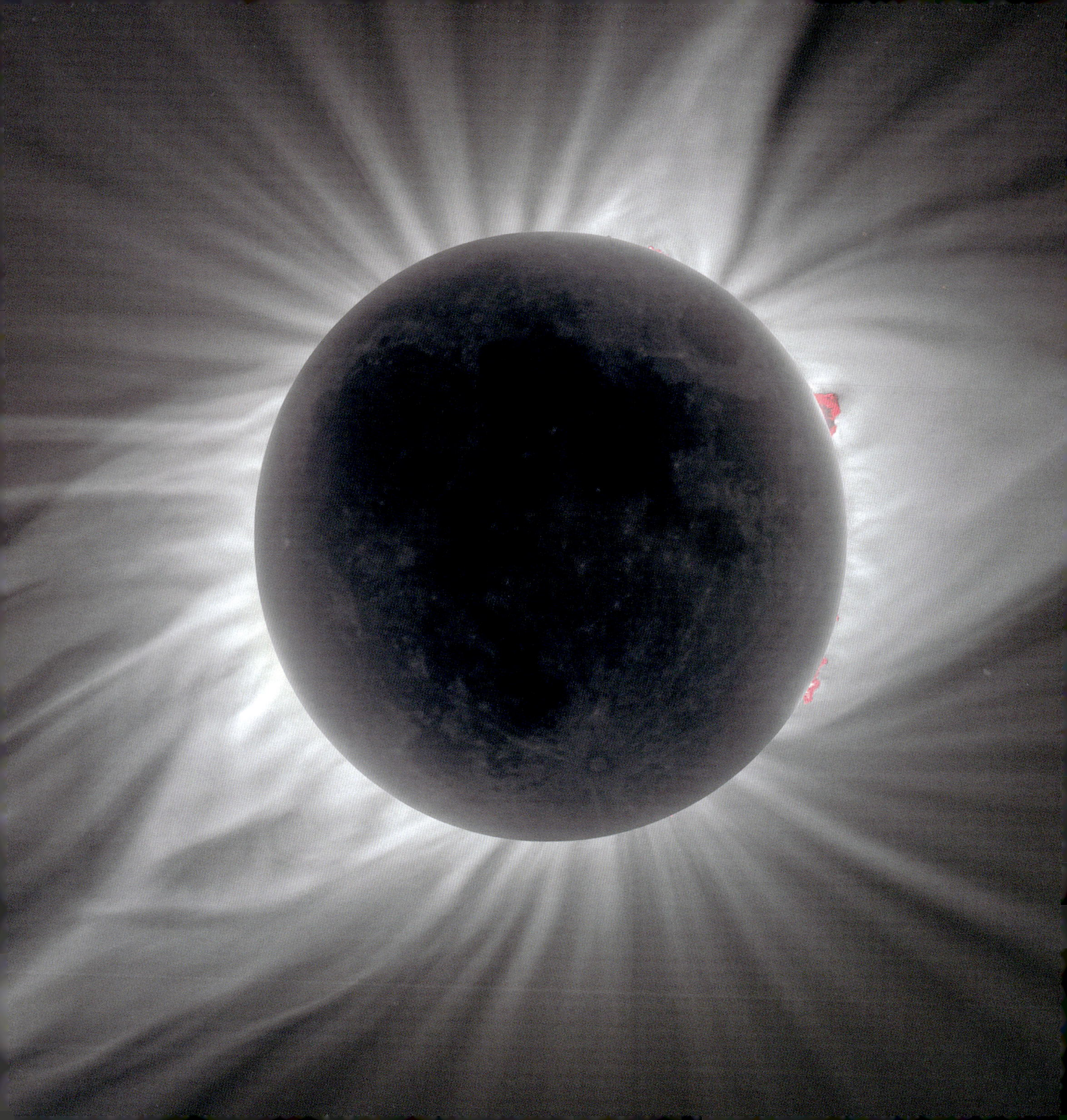

5

ECLIPSED

Eclipses are moments of high drama, a rare alignment of the Earth, Moon and Sun that briefly turns day into night. For many ancient civilisations they were portents of doom, heralding invasion, the fall of an empire or even the end of the world. Although science has long allowed us to predict their occurrence, eclipses retain an almost supernatural aura and thousands of people will travel from all over the world to stand under eclipse paths and watch as the Moon, for just a moment, blocks out the light of the Sun.

But eclipses are more than just spectacle. They reveal details of our nearest star that are otherwise hidden behind its dazzling light, from the tenuous corona to vast eruptions of superheated plasma. Over centuries, people have attempted to capture eclipses in a variety of media, and in doing so have uncovered many of the Sun's best-held secrets.

TABULAE ECLIPSIUM

Manuscript illumination on vellum by Joachinus de Gigantibus
for *Astronomia* by Christianus Prolianus, Naples, 1478
212 × 140 mm (manuscript page)
John Rylands Library

Eclipses have always been moments of both wonder and terror for human observers, despite our long-standing ability to predict their coming. Joachinus de Gigantibus produced a series of eight beautifully illuminated eclipse predictions for Christianus Prolianus's work on cosmology. He included both lunar and solar eclipses for the years 1468 to 1488 down the centre of the page reproduced opposite, with the years, months, dates and eclipse lengths listed alongside. The images show the Sun's or Moon's expected appearance at the maximum of the eclipse, with the Sun shining in gold leaf compared to the dull silver Moon. The manuscript later belonged to the Victorian designer William Morris, who used eclipses as a powerful literary image.

For more from *Astronomia*, see pp. 16–17.

Previous page
Detail from *The Big Corona* (see p. 115).

ãno·1486·	Erit eclipsis lu ne die·18·febr· hõ·0· minut 44	Durabit ista eclipsis per ho ras·3· minut ·44·
Eodē ãno·	Eclipsis solis die·4·marcij hõ·13· minut ·12·	Durabit ista eclipsis solis hõ·1 & minu ·40·
ãno·1487·	Lunae eclips die·7·februa rij hõ·10· minu ·23·	Tempus dura tionis eius e hõ·3· minut ·20·
Eodē ãno·	Erit eclipsis sol die·19· Iu lij·hõ·18·minu ·46·	Tempus dura tionis erit·1 hõ·0· minut ·40·
ãno·1488·	Eclipsis lunae Ianuarii die 28 hõ·2· minut 44	Durabit ista eclipsis lun hõ·0·& minu· ·14·
Eodē anno·	Erit eclipsis· solis die·8·Iu lij hõ·12·minu 4	Durabit ista eclipsis solis·1 hõ· ·minu

PLANISPHÆRII COELESTIS HEMISPHÆRIUM MERIDIONALE,
Calculatum ad finem Anni MDCC pro Ævo XVIII præsente: multis Stellis auctum et editum a CAROLO ALLARD, Amstelo-Batavo.
ex Officina I. COVENS ET C. MORTIER
Mercurius
Venus
Mars
Iupiter
Saturnus
Caliginum Lunæ
Motus Longitudinis Stellarum Fixarum, secundum Tychonem.
Motus Longitudinis Stellarum Fixarum, secundum Riccioli.
Planetarum a Terra Distantiæ Triplices
Planetarum Magnitudines e ratione Terræ.

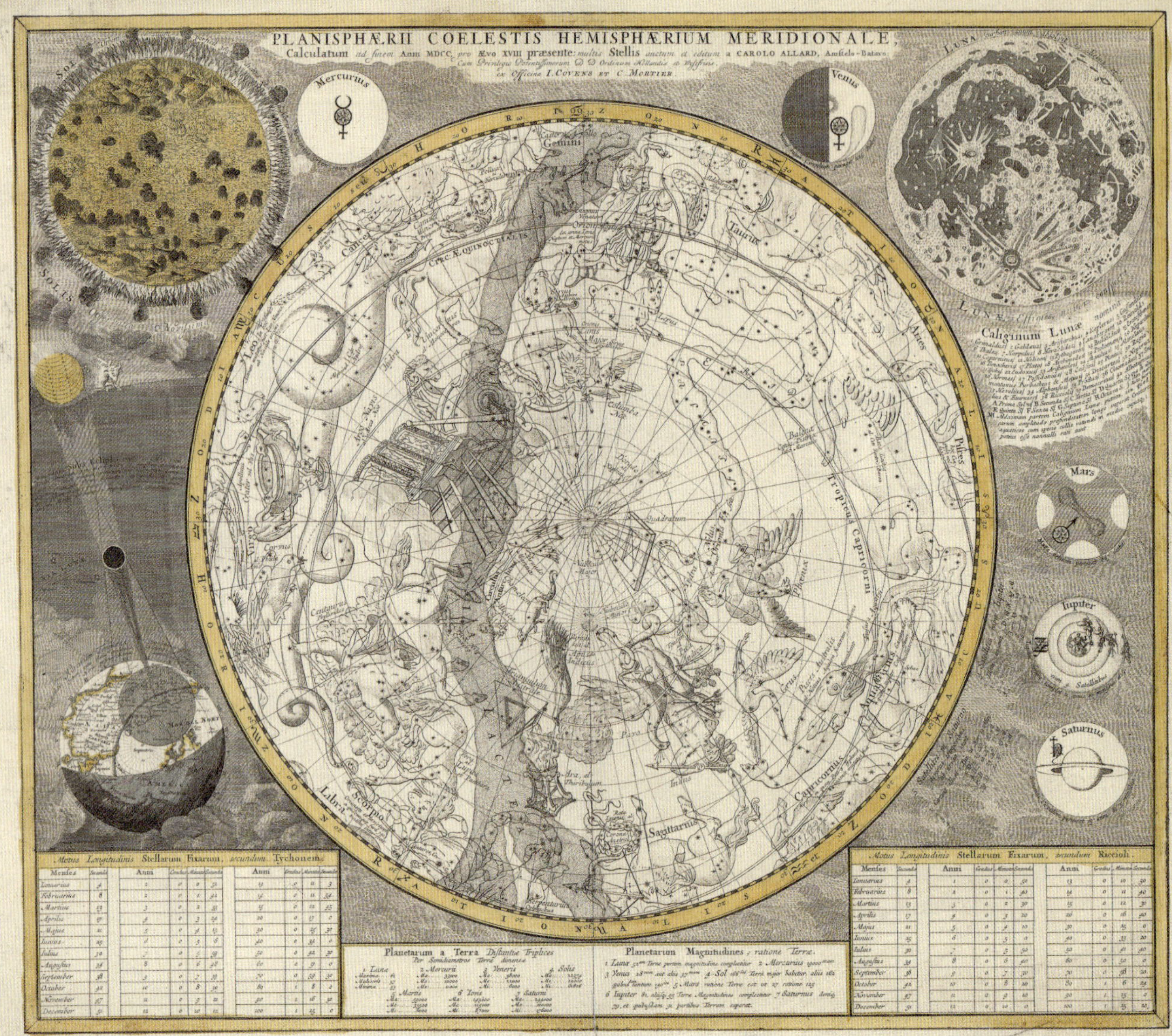

311A

PLANISPHAERII COELESTIS HEMISPHAERIUM MERIDIONALE: CALCULATUM AD FINEM ANNI MDCC, PRO AEVO XVIII PRAESENTE

Hand-coloured print by Carel Allard, Amsterdam, 1700
560 × 650 mm
Library of Congress

Carel Allard's richly detailed print represents a solar eclipse expected in 1706, but does so by focusing on the effects experienced on the Earth. The observer is put in an unusual viewing position above the Earth's North Pole in order to see the interplay of the diagrammatic Moon and Sun, which will cause darkness to fall over northern Europe. The viewer is encouraged to imagine how the Earth might look as a three-dimensional object from space. The diagram's eclipsed Sun is, likewise, placed directly beneath a representation of the Sun as a volcanic body copied from Athanasius Kircher, presenting it too as powerful and solid.

Kircher's volcanic Sun appears on pp. 64–5.

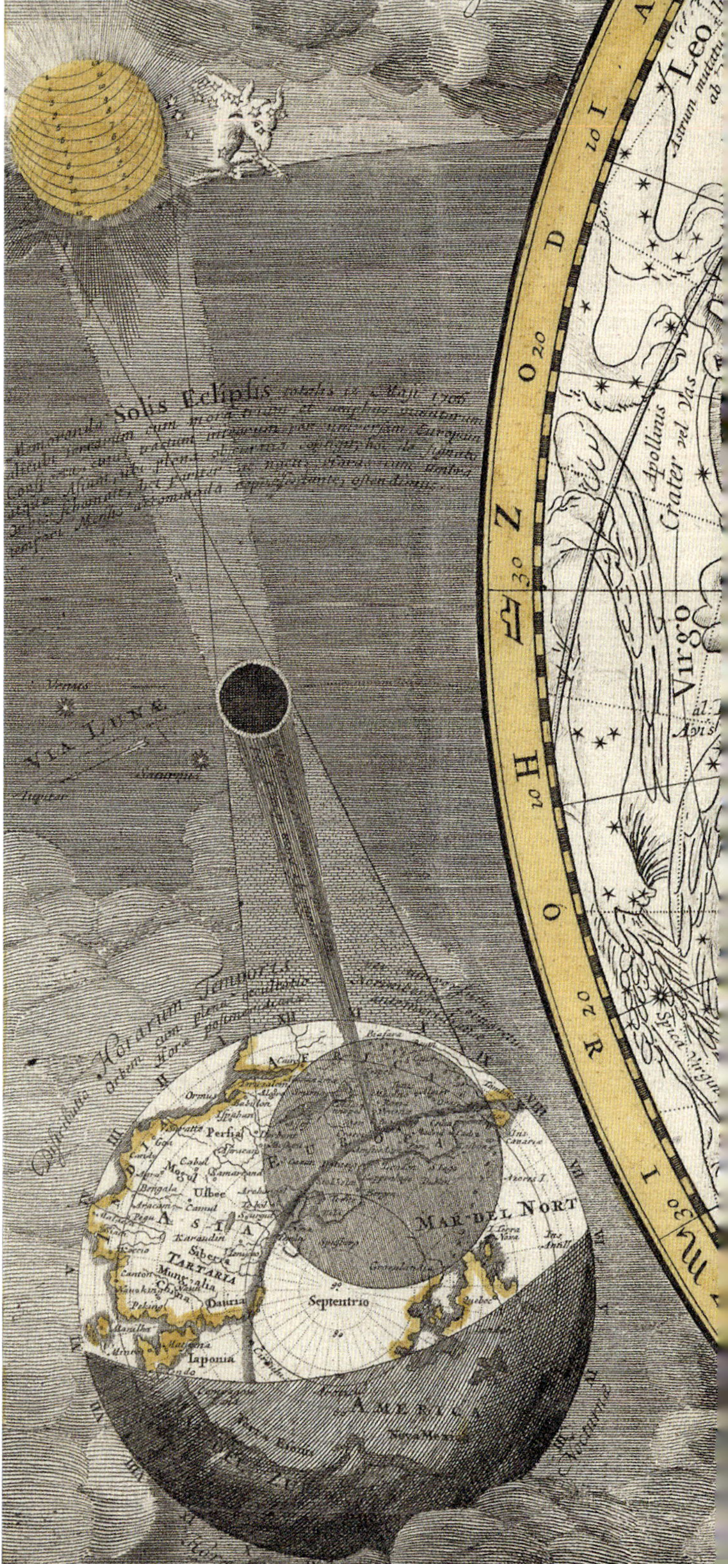

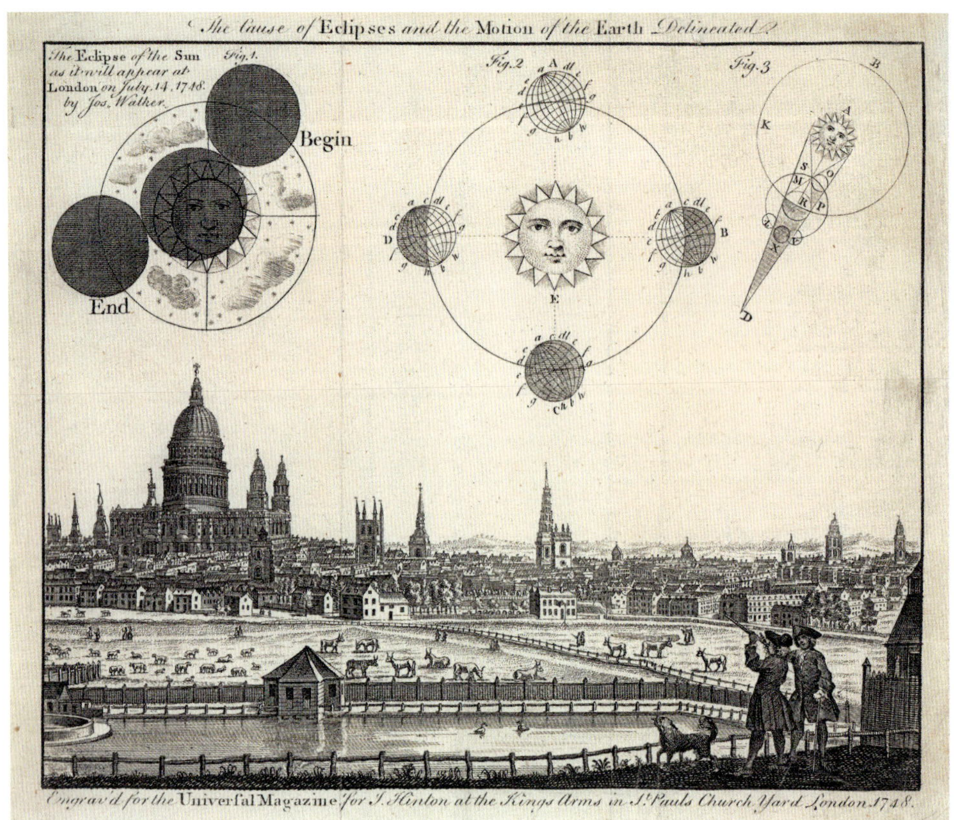

THE CAUSE OF ECLIPSES AND THE MOTION OF THE EARTH DELINEATED

Engraving by Joseph Walker published in *The Universal Magazine of Knowledge and Pleasure*, London, 1748
210 × 235 mm
Science Museum Group. Object no. 1999-1062

Astronomy and observational instruments were fashionable among the middle and upper classes in the 18th century. In 1748, *The Universal Magazine of Knowledge and Pleasure* printed this engraving demonstrating how an imminent partial solar eclipse would look from London on 14 July. Diagrams explaining the progress of the eclipse float in the sky over a detailed representation of London. Two men look across the Thames towards St Paul's Cathedral with their telescope. This is just one of a flood of eclipse images printed in London in the period, particularly surrounding the two total eclipses in 1715 and 1724. Other prints mapped the passage of the eclipse over the British Isles, or advertised lectures and instruments.

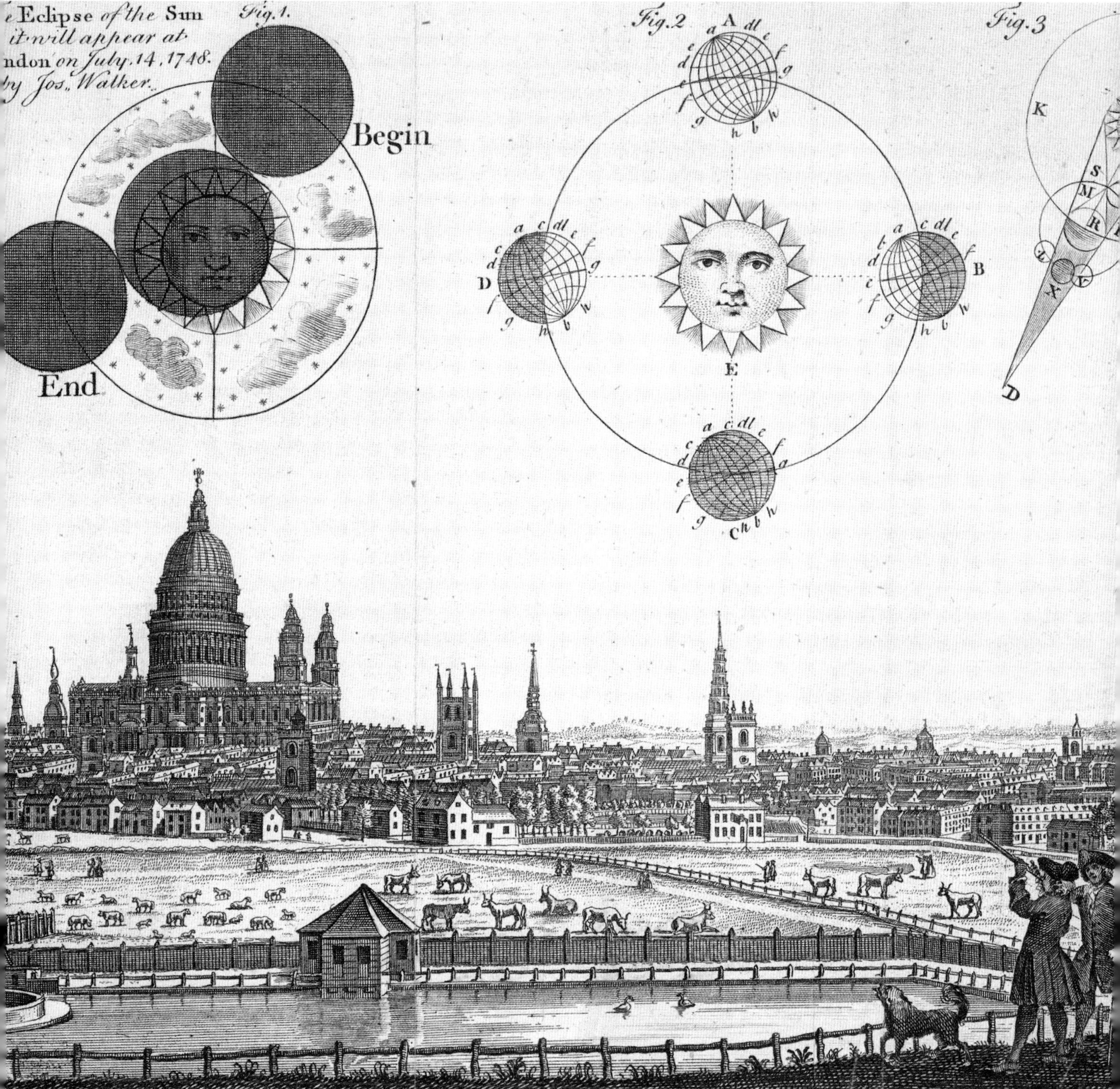

Eclipse of the Sun
it will appear at
ndon on July 14, 1748.
by Jos. Walker.
Fig. 1.
Begin
End
Fig. 2
Fig. 3

TOTAL ECLIPSE OF THE SUN, 22 DECEMBER 1870

Printed lithograph by Étienne Léopold Trouvelot for the *Annals of the Harvard College Observatory*, vol. 8, Cambridge, 1876
450 × 360 mm
Science Museum Group. Object no. 1887-23/7

Étienne Léopold Trouvelot created a number of classic images of the Sun at total eclipse. This one, produced for the *Annals of the Harvard College Observatory*, was specifically intended to show the corona, which is visible only when the Moon blocks the bright disc of the Sun. The lithographic process is particularly suited to portray the bright, hazy layers of the corona contrasted with the deep, inky eclipse. Trouvelot used photographs of the 1870 eclipse that captured the phenomenon of Baily's Beads, where the Moon's rugged surface allows the Sun to shine through at some points, but not others, creating the appearance of a string of beads circling the Sun.

Prints from Trouvelot's *Annals* also appear on pp. 47, 66, 68 and 69.

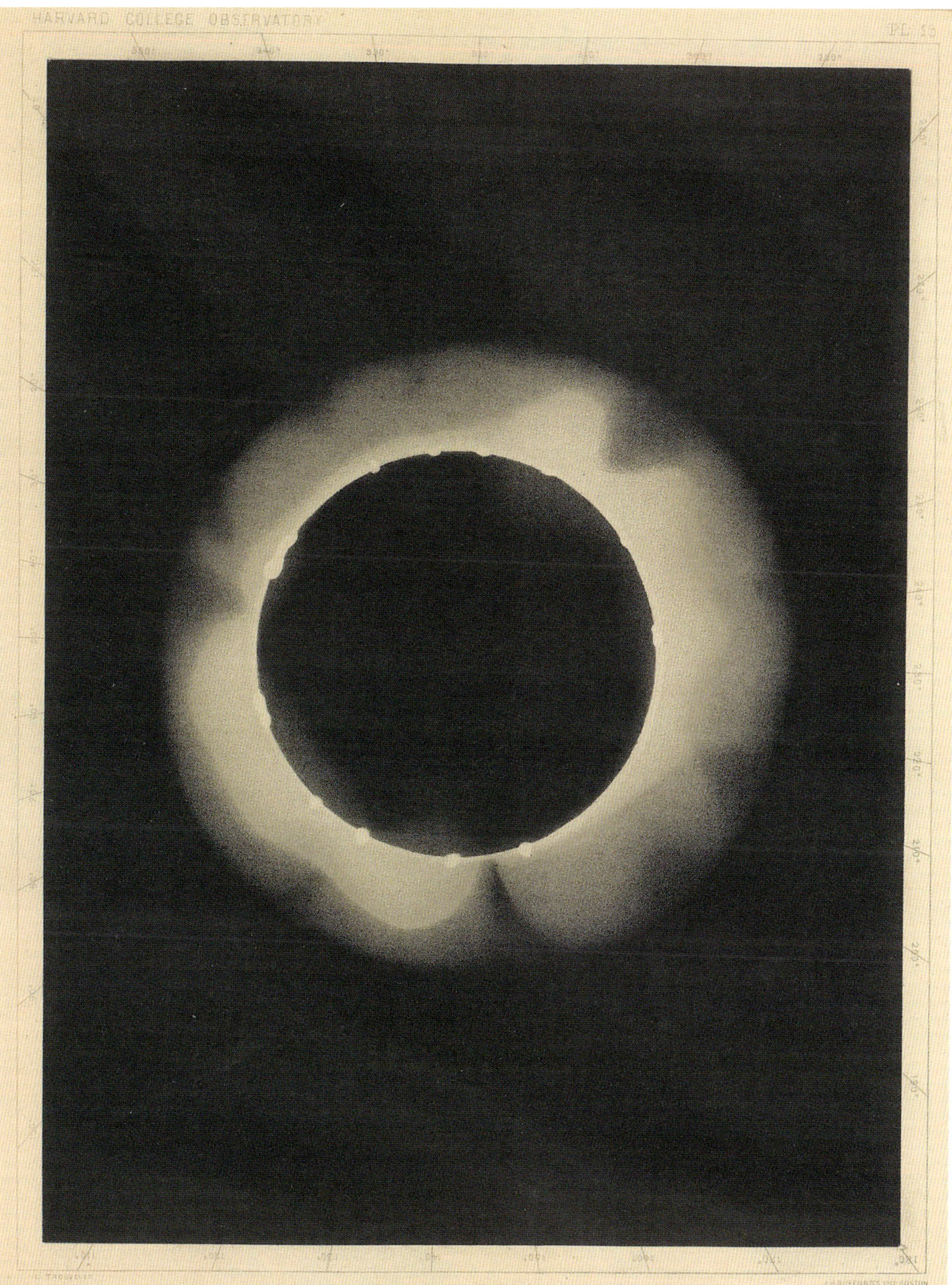
HARVARD COLLEGE OBSERVATORY
PL. 13
L. TROUVELOT
J. H. BUFFORD'S LITH. BOSTON

THE ECLIPSE OF 1860, BEFORE AND DURING TOTALITY

Glass positive photographs taken by Warren De La Rue using the Kew Photoheliograph from Rivabellosa, Spain, 18 July 1860
330 × 330 mm each
Science Museum Group. Object nos 1862-122/3 and 1862-122/4
Science Photo Library

These eclipse photographs were captured by Warren De La Rue using a telescope designed specifically for viewing the Sun, known as the Kew Photoheliograph. De La Rue took the instrument on an expedition to Spain in 1860 to obtain pictures of the eclipse on 18 July of that year.

Before photography, eclipse observations relied exclusively on the speed, skill and interpretation of an observer, leaving many questions unanswered about the nature of prominences and the corona. By systematically photographing the eclipse and comparing with the results of other observers, De La Rue confirmed them as features of the Sun, not the Moon or Earth.

Although the eclipse expedition was a success, De La Rue regretted that his scientific work prevented him from enjoying it fully, promising that next time he would devote himself 'to that full enjoyment of the spectacle which can only be obtained by the mere gazer'.*

* Warren De La Rue, 'The Total Solar Eclipse of 18 July 1860, observed at Rivabellosa, near Miranda De Ebro, in Spain', *Philosophical Transactions Part I* (London: Royal Society, 1862), p. 24.

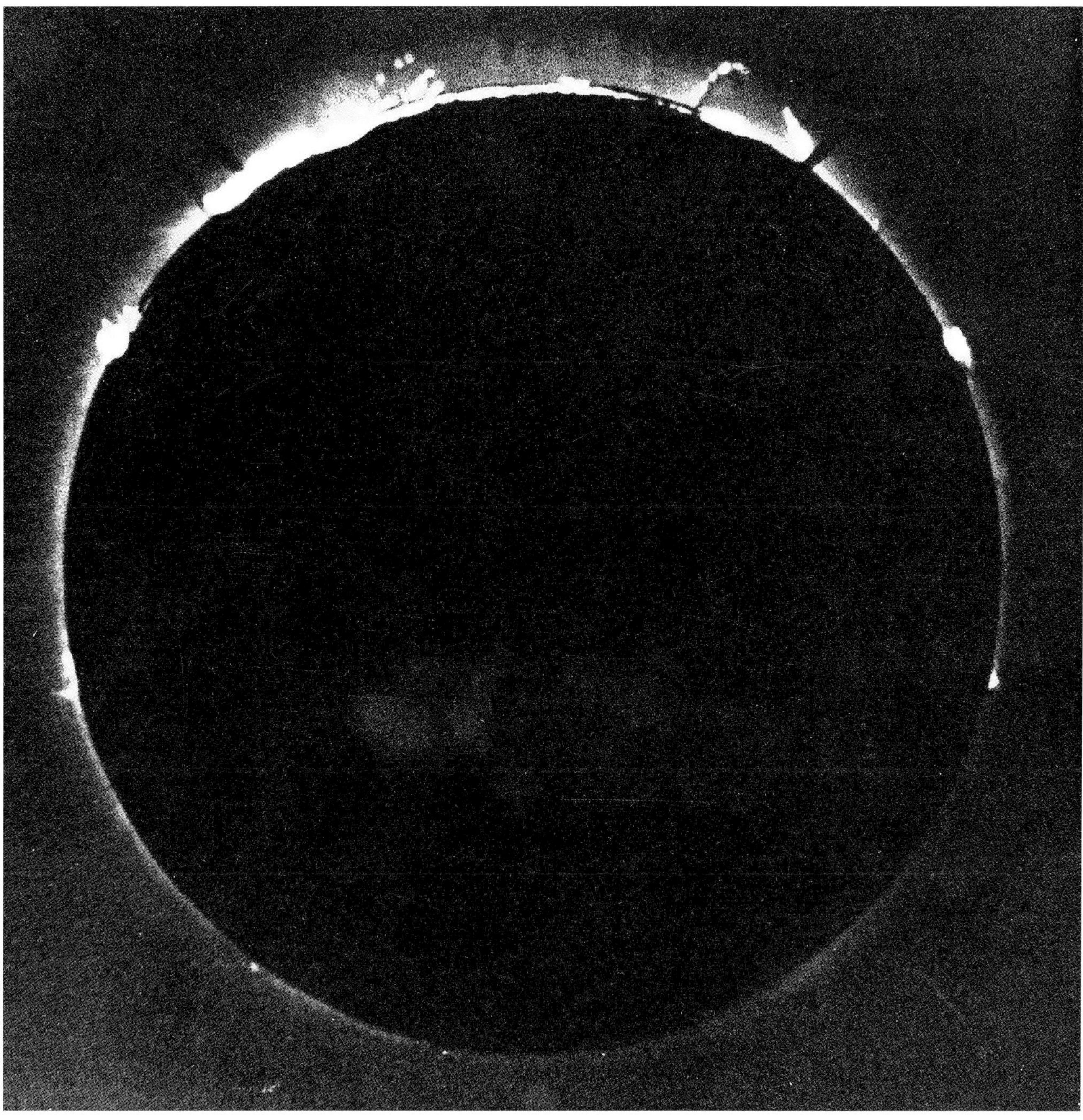

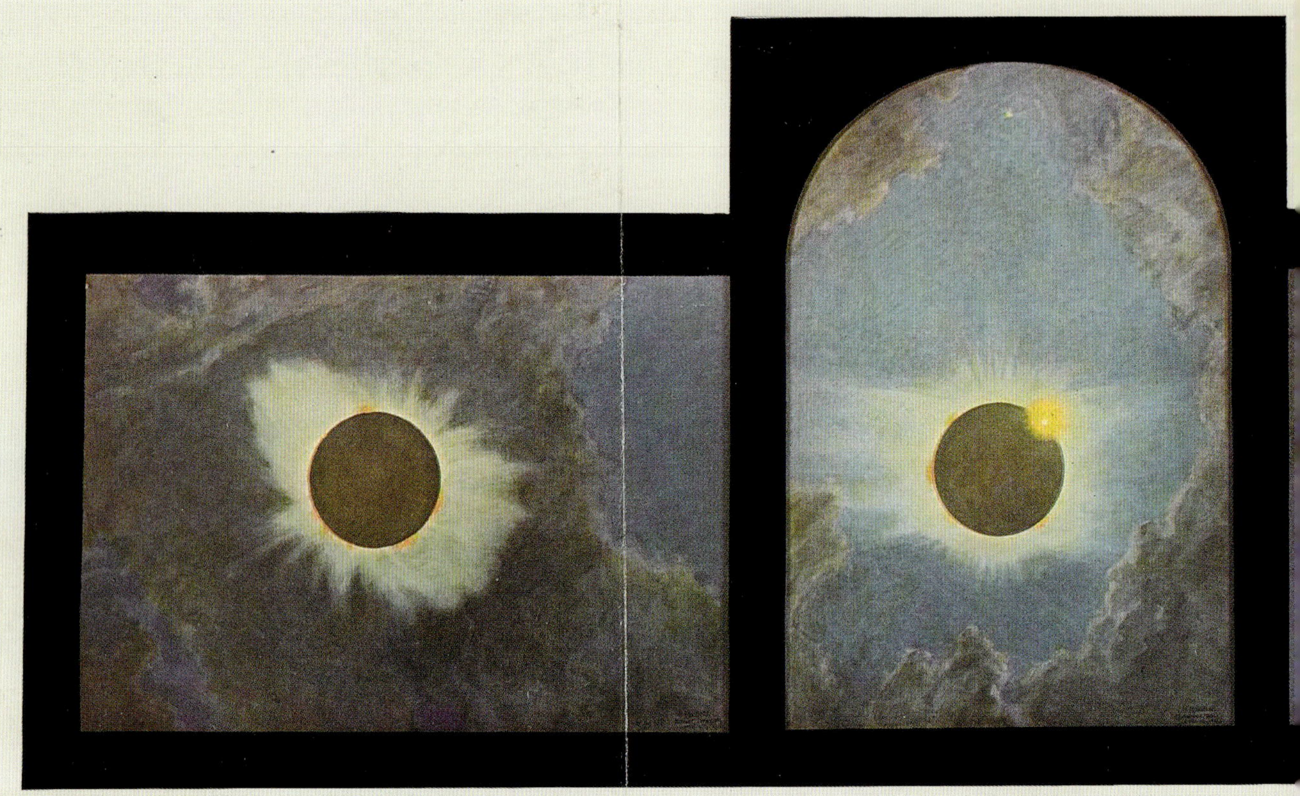

THE THREE SOLAR ECLIPSES SEEN IN THE UNITED STATES IN 1918,

FROM THE PAINTINGS BY HOWARD RUSSELL BUTLER

This series of eclipse paintings had its inception in the engagement of Mr. Butler by Mr. Edward Dean Adams to paint the eclipse of 1918, and Mr. Adams' subsequent gift of the pe
Proposed Hall of Astronomy, plans of which are now exhibited for the first time, and represents the first example of American Museum n

THE OREGON ECLIPSE, 1918

Totality Lasting 112 Seconds

As seen by Mr. Howard Russell Butler at Baker, Oregon, on June 8, 1918. The eclipse reached totality at 4:03 P.M. at an altitude of about 45°, and since it was in the afternoon, the long axis of the corona was inclined to the right. The time was near the period of maximum sun-spots, and as was to be expected, there was a diminished corona, but with more polar streamers than were expected. The prominences reached exceptionally large proportions, the "Heliosaurus" measuring 47,000 miles in height.

Gift of Mr. Edward Dean Adams to the American Museum of Natural History

THE CALIFORNIA ECLIPSE, 1923

Totality Lasting 140 Seconds

As seen near Lompoc, California, on September 10, 1923. This eclipse reached totality at 12:59 P.M., and sin
it was almost a noon-day eclipse, the long axis of the corona was practically horizontal. The time was nearer t
period of minimum sun-spots, and as was to be expected, the corona was more extended

Mr. Butler's painting shows the first Baily's bead of the third contact, that is, the first appearance of a spe
of the photosphere, evidently between two volcanic peaks on the rough surface of the moon.

Gift of the Morris K. Jesup Fund to the American Museum of Natural Histo

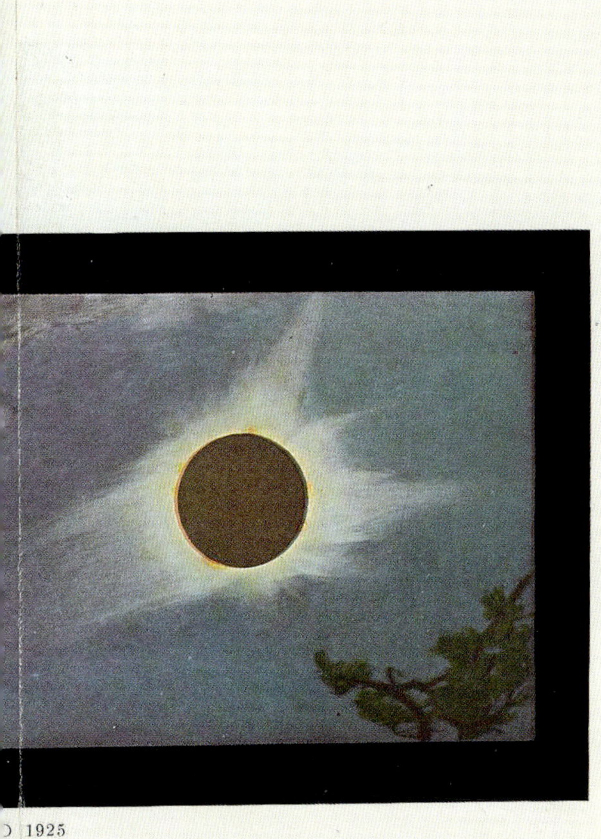

) 1925

useum. The present Triptych has been designed by Mr. Butler for the
tion applied to Astronomy.

THE CONNECTICUT-NEW YORK ECLIPSE, 1925

Totality Lasting 110 Seconds

As seen at Middletown, Connecticut, on January 24, 1925. This eclipse reached totality at 9:12 A.M., and con-
ently the long axis of the corona was inclined to the left. The time was near the period of maximum sun-spots
not only was the outline of the corona surprising to astronomers, but some of the streamers were much longer
was expected. The prominences, while many (19 were recorded by camera), were so small that they could
be seen by the naked eye.

Gift of the Morris K. Jesup Fund to the American Museum of Natural History

THE THREE SOLAR ECLIPSES SEEN IN THE UNITED STATES IN 1918, 1923 AND 1925. FROM THE PAINTINGS BY HOWARD RUSSELL BUTLER

Printed colour magazine supplement from *Natural History*, vol. 26, issue 4, New York, July 1926
232 × 436 mm
Science Museum Group. Object no. 1999-924

The triptych of paintings reproduced in this supplement combines the first works by a painter to capture accurately the solar corona. Howard Russell Butler trained as both a painter and physicist, and was invited to observe three eclipses with the US Naval Observatory. He developed a special system of note-taking in order to catch details of colour and size during the short-lived eclipse, and a schedule with ten-minute intervals called by an officer to keep his recording on track. He then painted immediately from his notes, photographs and remembered impressions. This magazine supplement announced (unfulfilled) plans for a Hall of Astronomy at the American Museum of Natural History, which Butler designed to house his huge triptych that measured over 6 metres wide.

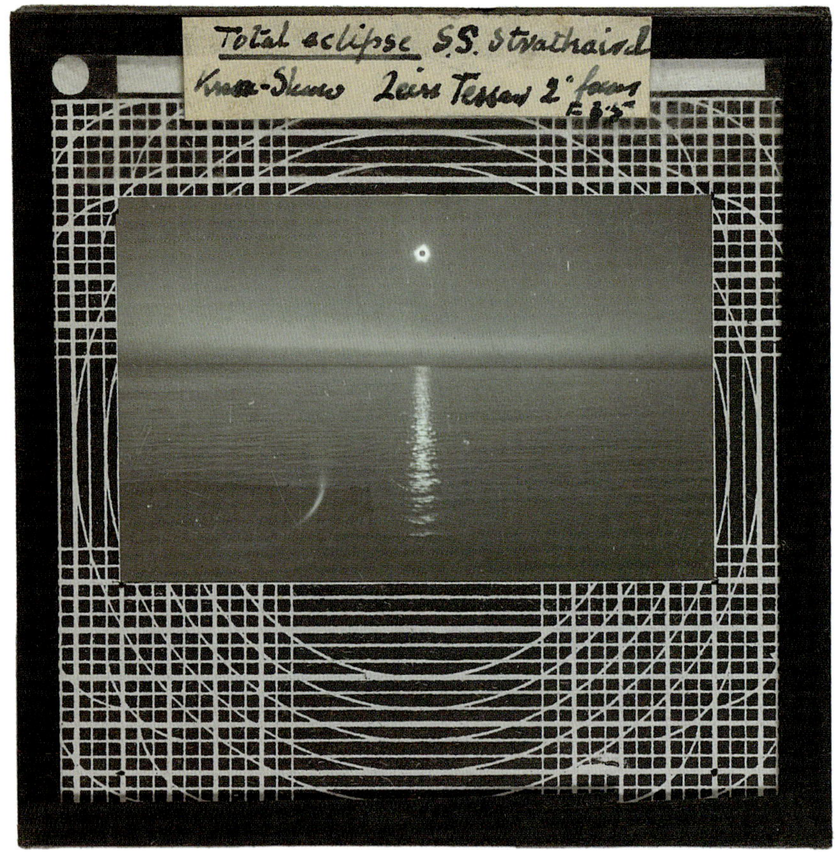

TOTAL ECLIPSE OVER THE AEGEAN SEA

Magic lantern slide. Photograph taken by John and Mary Evershed, Chios, Greece, 19 June 1936
83 × 83 mm
Science Museum Group. Object no. EVER/A/1/211

This photograph was taken by the astronomers John and Mary Evershed at the climax of an expedition to the Greek island of Chios to view the solar eclipse of 1936. Totality occurred early in the morning when the Sun was low to the horizon, with the corona reflected in the rippling waters of the Aegean. Unusually, the photograph sets the otherworldly eclipse against a relatable, even romantic, landscape: a harbour jetty, the sea and the sky. The Eversheds had travelled from Britain to Chios aboard the steamer RMS *Strathaird*, taking numerous photographs of their voyage and the eclipse itself. They had met 40 years earlier during an eclipse expedition to Norway, bonding over their love of astronomy and later marrying. The eclipse expedition of 1936 was one of several that they made together.

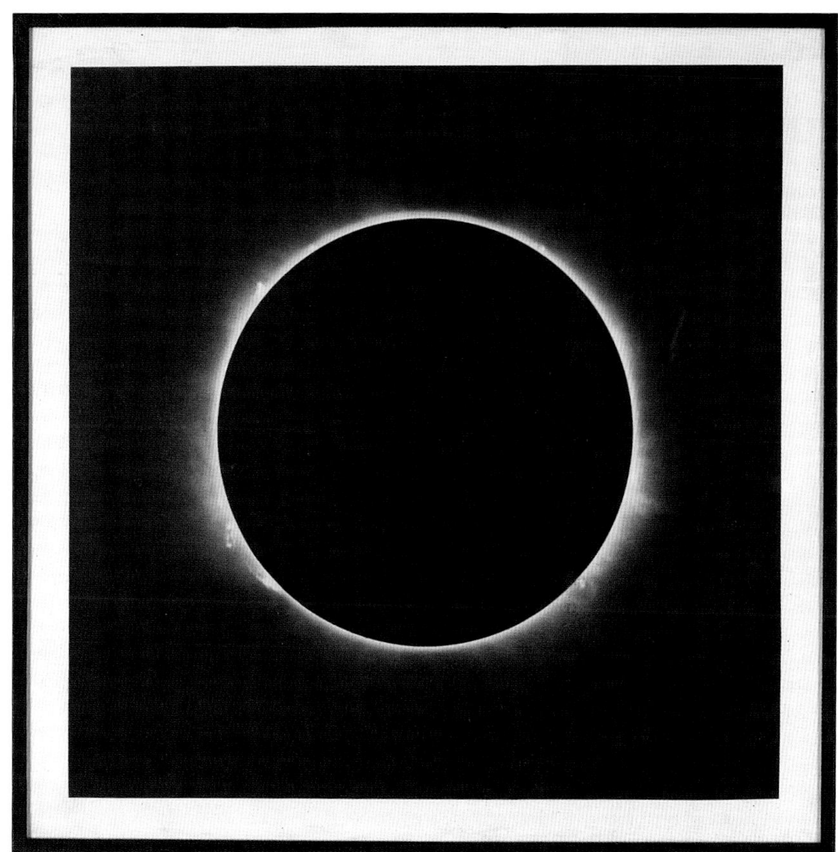

ARTIFICIAL ECLIPSE

Photograph of the solar corona taken with a coronagraph,
Pic du Midi Observatory, La Mongie, France, 7 July 1936
285 × 285 mm
Science Museum Group. Object no. 1938-217/3

This image may look like an eclipse, but appearances can be deceiving. In 1931 the French astronomer Bernard Lyot invented the coronagraph, a revolutionary instrument designed to produce an artificial eclipse so that the Sun's corona may be viewed at any time. The corona is around a million times fainter than the disc of the Sun, and is normally visible only when the brightness of the Sun is blocked by the Moon. A coronagraph removes light from the Sun's disc as well as light scattered by the Earth's atmosphere – the blue of the sky – which is itself strong enough to overwhelm the corona.

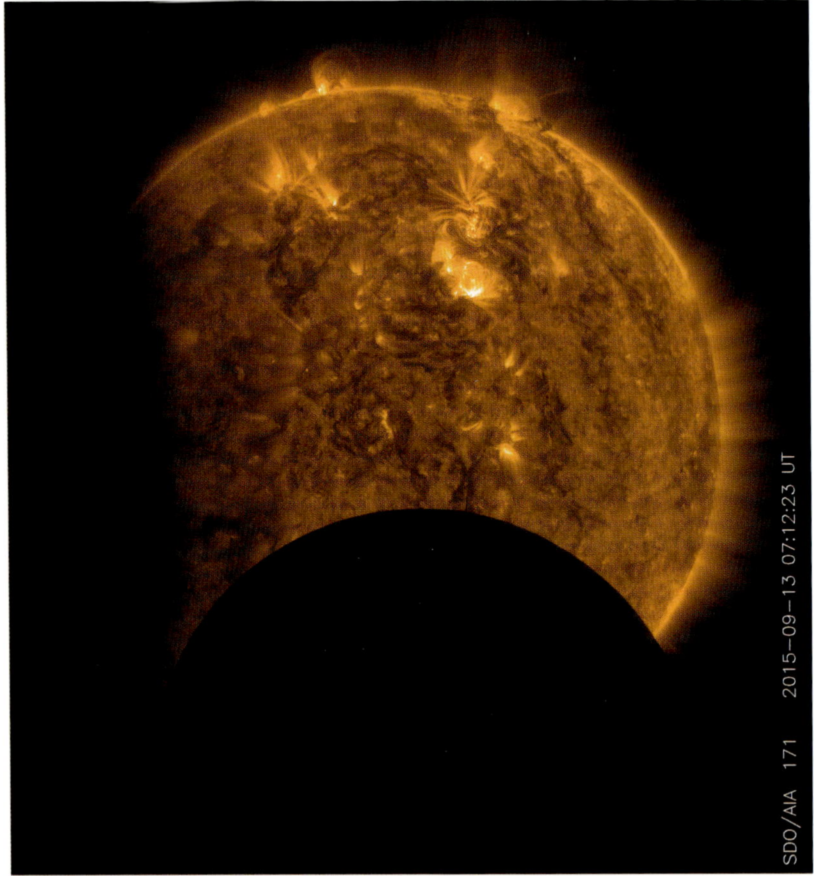

DOUBLE ECLIPSE

Photograph by the Solar Dynamics Observatory (SDO), 13 September 2015
Digital image
NASA

On 13 September 2015 NASA's orbiting Solar Dynamics Observatory witnessed an awesome astronomical alignment: a rare double eclipse of the Sun by the Moon and the Earth.The Moon appears as a crisp shadow at the bottom of the image, with the Earth on the left. The edge of the Earth looks fuzzy due to its thick atmosphere, which partially blocks the light of the Sun. The spectacle of a double eclipse is visible only from spacecraft orbiting the Earth.

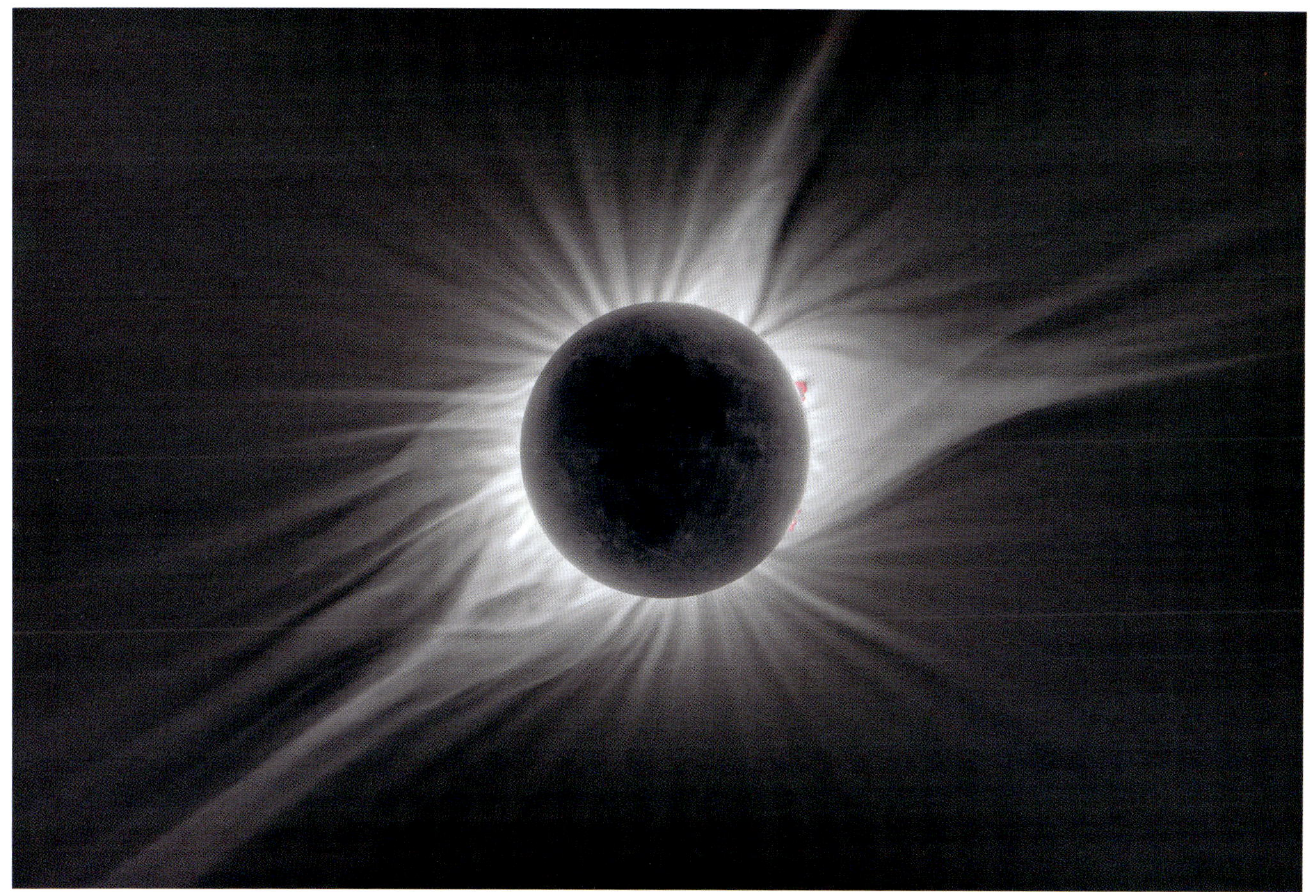

THE BIG CORONA

Composite photograph by Alson Wong, Jackson, Wyoming, USA, 21 August 2017
Digital image
Alson Wong

There is of course no substitute for experiencing an eclipse in person: cameras struggle to capture the full magnificence of the Sun's corona as viewed by the human eye. Yet digital photography can now produce startlingly beautiful images that show details we could never ordinarily see. The amateur astronomer Alson Wong digitally combined over 40 photographs with varying exposures to reveal the corona in all its glory: sheets of luminous gas and plasma looped by magnetic fields, streaming outwards from the solar surface. Close to the edge of the Sun, prominences stand out, bright pink against the corona. Even the face of the Moon is visible, illuminated faintly by light reflected from the Earth.

THE SHADOW OF THE MOON

Photograph by the Deep Space Climate Observatory, 21 August 2017
Digital image
NASA

This beautiful image of our planet was captured by NASA's Deep Space Climate Observatory during what was dubbed the 'Great American Eclipse' of August 2017. Reminiscent of many earlier eclipse maps, we see the shadow of the Moon over the Midwest as it makes its long journey across the entire United States, from the Pacific Northwest to the Atlantic Ocean.

On the ground, day turned briefly into an eerie twilight and observers were treated to a magnificent view of the solar corona. The eclipse of 2017 may have been the most watched in history: around 12 million people lived under the path of the Moon's shadow, with millions more travelling to the line of totality to witness one of the greatest spectacles that nature has to offer.

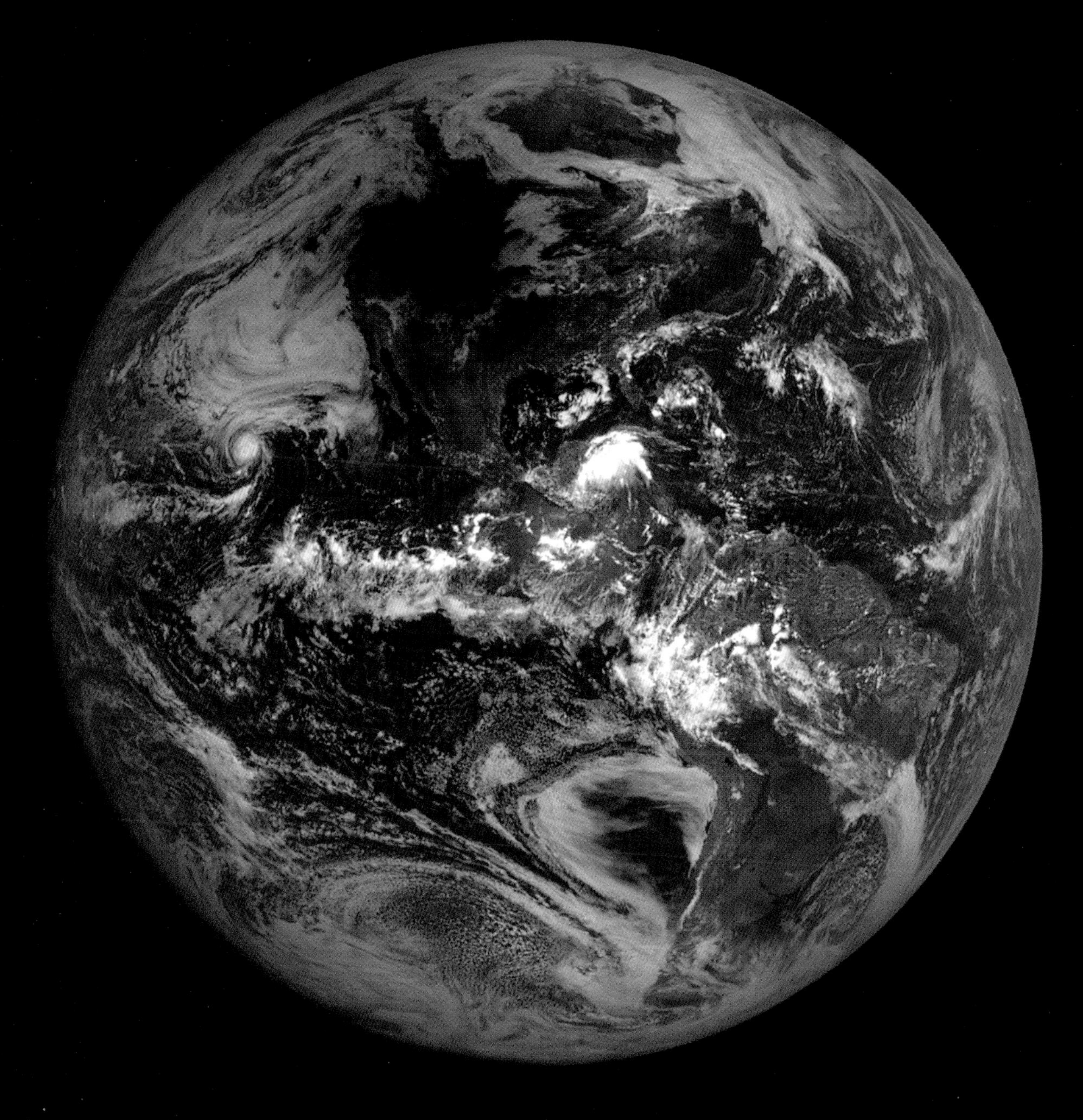

GLOSSARY

CHROMOLITHOGRAPH
Lithography is a means of producing a printed image directly from a freely drawn surface. The design is drawn onto a stone with a greasy substance, while the rest of the stone is treated to make it attractive to water. Ink made with oil will then stick to the drawn areas while being repelled by the others. Chromolithography allows an image to be printed in colour, using a different stone for each colour and repeating the printing process to layer the colours onto the paper.

CHROMOSPHERE
A brightly coloured layer of the solar atmosphere extending up to 5000 km above the visible surface, or photosphere. The rosy pink of the corona is mostly due to the emission of red light by superheated atoms of hydrogen. The chromosphere can be glimpsed briefly at the very start and end of totality during a solar eclipse.

CORONA
A tenuous halo of plasma surrounding the Sun, extending millions of kilometres into space. The corona is a trillion times less dense than the visible disc of the Sun and a million times fainter. It can be seen during total solar eclipses or imaged using specially designed instruments.

CORONAGRAPH
An instrument that can be attached to a telescope in order to view the solar corona at any time when the Sun is in the sky. It produces an artificial eclipse by blocking or removing the light from the visible disc of the Sun and uses polarising filters to remove light scattered by the Earth's atmosphere, which would otherwise overwhelm the faint light of the corona.

CORONAL MASS EJECTION (CME)
A large eruption of plasma from the Sun's corona that escapes the Sun's gravitational pull and travels outwards through the solar system. CMEs are thought to be triggered by magnetic fields close to the Sun's surface undergoing a violent rearrangement. Large CMEs can present a significant hazard to human technology, with the potential to disrupt or damage communication networks, electricity supplies and orbiting satellites and spacecraft.

DAGUERREOTYPE
The first publicly available form of photography, invented by the French artist Louis-Jacques-Mandé Daguerre. Sheets of silver-plated copper were treated with vapour of iodine, bromine and/or chlorine before being exposed to light in a camera. Mercury vapour was then used to develop the photographic image, which was sealed behind glass to protect its fragile surface.

ELECTROMAGNETIC RADIATION
A form of radiation composed of oscillating electric and magnetic fields that travels at the speed of light in a vacuum. Visible light, radio waves, ultraviolet and X-rays are all forms of electromagnetic radiation.

ENGRAVING
A means of producing a printed image. The engraver uses a sharp tool called a burin to gouge a design into a copper plate. Ink is then worked into the engraved lines, and the plate is run through a press with a piece of paper. The image is produced in reverse where the ink lines pass to the paper.

ETCHING
Can be used in conjunction with engraving. Here the copper plate is coated in a waxy substance that is resistant to acid. The engraver then draws a design into the soft wax, exposing the metal surface. Next, the plate is put into a bath of acid, which etches into the plate where the design is exposed, leaving unaffected the parts coated in wax.

FLARE
A sudden, extremely powerful flash of light from the Sun, across all wavelengths from radio waves to gamma rays. Flares usually last for only a few minutes and often accompany a release of plasma from the corona, known as a coronal mass ejection. The intense burst of ultraviolet light and X-rays released by flares can disrupt radio communication on Earth.

HELIOSCOPE
A method whereby a telescope is employed to project an image of the Sun onto a screen or sheet of paper, so that it can be studied and sketched exactly. Helioscopes were first used in the 17th century to draw sunspots on the surface of the Sun.

ILLUMINATED MANUSCRIPT
Manuscripts were the primary means for the circulation of images until the invention of the printing press in the 15th century. They were produced by hand in ink on vellum, parchment or paper. Illumination was the process of adding images and embellishments in the form of illustrations, marginalia (borders) or historiated initials (enlarged, decorated capitals at the start of a paragraph).

NEGATIVE
A negative is a photographic image in which the light areas appear dark and the dark areas light. It is usually the first stage in the production of a positive photograph and is created using light-sensitive chemicals.

PENUMBRA
The brighter outer edge of a sunspot surrounding the darker central umbra.

PHOTOGRAPHY
Photography is any process that creates an image on a light-sensitive surface. Initially photographs were produced on metal, glass or paper using chemicals, but today we also apply the term to images taken with digital technology.

PHOTOHELIOGRAPH
A specially designed telescope used to take photographs of the Sun, invented in the 1850s.

PHOTOSPHERE
The visible surface of the Sun, with a temperature of around 5500 degrees Celsius. The photosphere lies just below the chromosphere and is the location of sunspots.

PLASMA
The fourth state of matter beyond gas, where high temperatures cause molecules and atoms to disintegrate to form an electrically charged gas of electrons and ions. The Sun is an almost perfect sphere of plasma, which is also the most abundant state of ordinary matter in the universe.

POSITIVE
A positive is the image from a photograph that records the relative lights, darks and colours as they appear to the naked eye.

PROMINENCE
A large bright feature extending outwards from the visible surface of the Sun. Prominences are made of plasma and can rise hundreds of thousands of kilometres above the solar surface. Astronomers have also called these 'protuberances' and 'eruptions'.

SOLAR WIND
A stream of plasma, flowing outwards from the Sun towards the edge of the solar system. Originating in the solar corona, the solar wind is mostly made of electrons, protons and helium nuclei, and carries magnetic fields with it as it travels outwards from the Sun.

SPECTRAL LINE
Spectral lines come in two types. 'Absorption' lines appear as dark lines in a spectrum of light, and are caused by certain atoms absorbing particular colours (wavelengths) of light. 'Emission' lines appear as bright lines, and are caused by the emission of certain colours of light by atoms when they are heated to high temperatures. The colours that a particular atom absorbs are identical to those it emits.

SPECTROGRAPH
An instrument used to break sunlight into its spectrum so that it can be studied in detail. Spectrographs usually use multiple prisms to spread the spectrum over as large an angle as possible so that fine details in the spectral lines are revealed.

SPECTROHELIOGRAPH
An instrument designed to image the Sun in a single colour or wavelength of light or electromagnetic radiation.

SPECTROSCOPY
The study of the spectrum of light, particularly spectral lines. Spectroscopy can be used to identify the make-up of an astronomical body, such as the Sun, by comparing the position of its spectral lines with those obtained from known chemical elements in a laboratory.

SPECTRUM
The band of colours produced when light is dispersed into its component colours, for example, using a prism. In the case of visible sunlight, this appears as a rainbow of colour running from red to violet.

SUNSPOT
A dark spot on the visible surface of the Sun. Sunspots shine brilliantly, but look dark due to their low temperature relative to the rest of the solar surface. They appear where magnetic loops generated inside the Sun break through the solar surface and prevent upwelling hot plasma from reaching the surface. They are often the source of solar flares and coronal mass ejections.

TELESCOPE
An instrument used to magnify distant objects. Probably invented at the start of the 17th century by Dutch opticians, telescopes were quickly applied to studying the heavens, leading to numerous discoveries including the moons of Jupiter, the phases of Venus and sunspots. Optical telescopes fall into two broad categories: refracting (gathering light with lenses) or reflecting (using mirrors). The word 'telescope' is also often employed for detectors exploring different parts of the electromagnetic spectrum.

ULTRAVIOLET
A form of electromagnetic radiation lying beyond the violet end of the visible spectrum. The Sun shines brightly in ultraviolet light, but this is mostly absorbed by the Earth's atmosphere. Orbiting spacecraft are able to image the Sun in ultraviolet, revealing features that are not apparent in visible light alone.

UMBRA
The dark centre of a sunspot, where the magnetic field emerges from within the body of the Sun.

FURTHER READING

Universe: Exploring the Astronomical World (London: Phaidon Press Ltd, 2017).

Michael Benson, *Cosmigraphics: Picturing Space through Time* (New York: Abrams, 2014).

Lorraine Daston and Peter Galison, *Objectivity* (New York: Zone Books, 2007).

Leon Golub and Jay M Pasachoff, *The Sun* (London: Reaktion Books, in association with the Science Museum, 2017).

Lucie Green, *15 Million Degrees: A Journey to the Centre of the Sun* (London: Penguin, 2017).

Klaus Hentschel, *Mapping the Spectrum: Techniques of Visual Representation in Research and Teaching* (Oxford: Oxford University Press, 2002).

Nick Kanas, *Star Maps: History, Artistry, and Cartography* (Berlin: Springer in association with Praxis Publishing, 2007).

Nick Kanas, *Solar System Maps: From Antiquity to the Space Age* (New York: Springer in association with Praxis Publishing, 2014).

Omar W Nasim, *Observing by Hand: Sketching the Nebulae in the Nineteenth Century* (Chicago: University of Chicago Press, 2013).

ACKNOWLEDGEMENTS

Just like the images and texts that are our subject, this book has been an inherently collaborative effort.

We would like to thank our colleagues in the Science Museum Group who have helped in the book's production: Wendy Burford and Charlotte Grievson in Publishing; John Herrick, Jennie Hills and Kira Zumkley in Photography; and Beata Bradford, Jessica Crann, Prabha Shah, Doug Stimson, John Underwood, Ian Wilkes and Nick Wyatt in the Library and Archives.

Colleagues in the Science Museum Group and elsewhere have likewise kindly offered their advice and expertise on the images featured. Our thanks to Geoff Belknap, Jim Bennett, Alison Boyle, Oliver Carpenter, Louise Devoy, Seb Falk, Lucie Green, Emma Hedderwick, David Rooney and Melanie Vandenbrouck.

PICTURE CREDITS